Springer Theses

Recognizing Outstanding Ph.D. Research

Aims and Scope

The series "Springer Theses" brings together a selection of the very best Ph.D. theses from around the world and across the physical sciences. Nominated and endorsed by two recognized specialists, each published volume has been selected for its scientific excellence and the high impact of its contents for the pertinent field of research. For greater accessibility to non-specialists, the published versions include an extended introduction, as well as a foreword by the student's supervisor explaining the special relevance of the work for the field. As a whole, the series will provide a valuable resource both for newcomers to the research fields described, and for other scientists seeking detailed background information on special questions. Finally, it provides an accredited documentation of the valuable contributions made by today's younger generation of scientists.

Theses may be nominated for publication in this series by heads of department at internationally leading universities or institutes and should fulfill all of the following criteria

- They must be written in good English.
- The topic should fall within the confines of Chemistry, Physics, Earth Sciences, Engineering and related interdisciplinary fields such as Materials, Nanoscience, Chemical Engineering, Complex Systems and Biophysics.
- The work reported in the thesis must represent a significant scientific advance.
- If the thesis includes previously published material, permission to reproduce this must be gained from the respective copyright holder (a maximum 30% of the thesis should be a verbatim reproduction from the author's previous publications).
- They must have been examined and passed during the 12 months prior to nomination.
- Each thesis should include a foreword by the supervisor outlining the significance of its content.
- The theses should have a clearly defined structure including an introduction accessible to new PhD students and scientists not expert in the relevant field.

Indexed by zbMATH.

Francisco Restrepo

Angle-Resolved Photoemission Spectroscopy Study of Spin Fluctuations in the Cuprate Superconductors

Doctoral Thesis accepted by the University of Illinois at Chicago, USA

Springer

Francisco Restrepo
Department of Physics
University of Virginia
Charlottesville, VA, USA

ISSN 2190-5053 ISSN 2190-5061 (electronic)
Springer Theses
ISBN 978-3-031-10978-2 ISBN 978-3-031-10979-9 (eBook)
https://doi.org/10.1007/978-3-031-10979-9

This Springer imprint is published by the registered company Springer Nature Switzerland AG
The registered company address is: Gewerbestrasse 11, 6330 Cham, Switzerland

To my parents, Rosa Helena Medina and Juan Guillermo Restrepo

Supervisor's Foreword

Shortly after the discovery of the lanthanum barium copper oxide high-temperature superconductors, P.W. Anderson pointed out that the parent insulator was an odd electron state with unusual magnetic properties, an antiferromagnetic Mott insulator. Several authors considered the possibility that as the insulator is doped, antiferromagnetic fluctuations, generically referred to as spin fluctuations, persisted in the doped insulator. It was suggested that these spin fluctuations could effectively become the glue that binds the electrons in the Fermi sea, leading to high-temperature superconductivity. The transition temperature T_c could be calculated if one knew the spin fluctuation spectra, the electronic spectra, and the interaction between the two.

Calculations, however, showed mixed results: some indicating that high T_c could originate from these interactions, while others did not produce reasonable values of T_c. There was also controversy over the pseudogap, the state above T_c in the optimally and underdoped materials. A large part of the problem was the fact that one of the components needed to calculate T_c was not known a priori. For example, the electron spectra of some materials were not measured at temperatures where the material would have become a superconductor, or the electron spectra would not have shown a superconducting gap at low temperatures, for example.

More recently, extensive data sets under far better controlled conditions became available. Dr. Restrepo used experimental techniques that allowed the samples to keep their surface cleanliness, and therefore their doping values, for over a month under experimental conditions. This is crucial in allowing measurements of the electronic excitation spectra of the entire irreducible Brillouin zone under identical conditions. He mastered the details of the entire measurement process to a point where the experiments provided very reliable data. From previous work, a general understanding developed, that in order to make a reasonable estimate of T_c, all of the quantities entering the calculations needed to be known in the normal state at a temperature close enough above T_c, and at the appropriate doping level, so as to lead to a reasonable estimate of T_c.

Perhaps a more serious problem for this entire program was the fact that the spin fluctuation spectra above T_c were not considered to be representative of those in the superconducting state. A few years before the present work, Chatterjee et al. had shown that, surprisingly, ARPES could be used to derive a very close approximation to the spin excitation spectra measured by neutron scattering methods, even though ARPES only measures the single-particle scattering component of the electron-spin excitation spectra, while the neutron scattering spectra include all orders of scattering. And a priori, one does not know if the higher order processes are critical in describing the important consequences of the electron-spin fluctuation scattering.

Nonetheless, Dr. Restrepo decided to test the spin fluctuation theory again, making use of the appropriate bosonic and fermionic functions obtained from ARPES data in the normal state, and at the correct temperatures. This was a remarkably brave attempt, considering that Dr. Restrepo was trained as an experimental physicist, while the calculations required some of the most advanced theoretical and calculational methods. For example, the strong interactions involved required the use of a generalization of the BCS gap equation –the Bethe–Salpeter equation (BSE)– to calculate T_c. To this end, Dr. Restrepo first showed how the various terms entering the BSE equation could be obtained, within certain approximations, from the spectrum of electronic excitations measured by ARPES. He then used these ARPES-derived quantities as inputs to the BSE and found solutions corresponding to T_c values of 60–90 K, in agreement with experiments. Thus, he concluded that antiferromagnetic correlations are a viable candidate for the pairing interaction in the cuprate superconductors.

Aside from carrying out the measurements, Dr. Restrepo himself also carried out all the calculations required for this work, a rare accomplishment for an experimentally trained physicist. Moreover, I should point out that previous to the measurements and analysis of the data, Dr. Restrepo was involved in the construction of the measurement apparatus. This entire body of work is described in detail in the PhD thesis. Chapter 1 introduces the problem of high T_c superconductivity and two of the main classes of theories proposed to solve it. One such class of theories, studied in this work, explores the possibility of magnetic excitations as mediators of Cooper pairing, as mentioned above. Chapter 2 summarizes the basic concepts of superconductivity in Green's function formalism and offers a working introduction to the cuprate superconductors and their anomalous characteristics. Chapter 3 introduces the experimental technique used in this work, ARPES. Chapter 4 covers some of the more technical aspects of the experiment, highlighting Dr. Restrepo's contributions to the construction of measurement instrument. In Chap. 5, Dr. Restrepo shows how the extensive ARPES data can be used to extract the single-particle Green's function, as well as construct the spin susceptibility, both of which are used as inputs to the Bethe–Salpeter equation. In doing so, he finds eigenvalues of this equation consistent with T_c values in agreement with experiments and eigenfunctions—which reflect the properties of the superconducting order parameter—with d-wave symmetry and an energy dependence consistent with the experimentally observed time retardation of the pairing interaction. Overall, his results represent an attempt to quantitatively reconcile observations from various experimental probes—like neutron scattering, ARPES, and others—and account for the high critical temperatures observed in the cuprate superconductors.

Dr. Restrepo has also striven to make his work self-contained and pedagogical, as evidenced by the extensive introductory material contained in Chaps. 1–4, as well as the additional explanations in the appendices. Therefore, we hope that new students in high T_c superconductivity and photoemission spectroscopy will find this work original, engaging, and instructive. We wish to thank Springer for giving us the opportunity to share this exciting work with a wider audience.

University of Illinois at Chicago
Chicago, IL, USA
May 2022

Juan Carlos Campuzano

Acknowledgments

I want to first thank my advisor, Juan Carlos Campuzano, for his understanding, his seemingly infinite patience, and, of course, his advice, which often went beyond the scientific. I am also indebted to professors Utpal Chatterjee and Dirk Morr for several key suggestions which helped this project come to fruition. (I am particularly grateful to Professor Chatterjee for kindly sharing a large data set analyzed in this work.) I must also thank my colleague, Hao Xu, for his perseverance and resourcefulness in solving the many technical difficulties that plagued our experiment. I am grateful to Fanny Rodolakis and Cris Adriano for helpful advice during the many hours spent at the synchrotron facility during the first two years of my research. It is also a great pleasure for me to acknowledge the outstanding work performed by the instrument makers at UIC; in particular, Kevin Lynch, Bob Kurdydyk, Rick Frueh, and Rich Djoutrek. The many parts they fabricated for us, their aid in the design process, and the instrumentation difficulties they helped us overcome are testimony to a contribution to this work that cannot be overstated. Also at UIC, I wish to thank the graduate student advisor James Nell for his patience and his eagerness to help, not just me, but also other physics graduate students. I am equally grateful to Derrick Stanley and Brian Shim for placing orders for me in a timely fashion and keeping our accounts under control. Professors Mark Schlossman and Anjum Ansari also deserve special mention. Aside from being outstanding physics instructors, they provided me with a better appreciation of the true nature of scientific research by kindly allowing me to work in their labs as an undergraduate assistant. I also want to thank my defense committee for their interest in my work and for making part of an important event in my career.

On a more personal side, I would like to extend my gratitude to my family, especially to my parents, for their unfaltering support and encouragement throughout my life. This work is theirs. Finally, I should very much like to thank the friends I made during my stay at UIC, as well as my old childhood friends from Colombia. It is no exaggeration to say that through all our bickering, arguing, joking, laughing, agreeing, and disagreeing, they have left their mark in my memory and thought.

Parts of This Thesis Have Been Published in the Following Journal Article

Coupled electronic and magnetic excitations in the cuprates and their role in the superconducting transition. *Commun. Phys.* **5**, 45 (2022)

Contents

Chapter 1
Introduction

Abstract The temperature versus doping phase diagram of the high-temperature cuprate superconductors exhibits several anomalous phases in conflict with the conventional band theory of solids. Two broad classes of theories attempting to explain these phases can be identified: the first one emphasizes the strong Coulomb repulsion in the undoped parent compound and the effect of adding charge carriers to such a correlated state; in the second class, the systems are described as Fermi liquids whose properties are strongly renormalized by magnetic excitations (or spin fluctuations). In the latter theories, it has been hypothesized that superconductivity itself may be mediated by spin fluctuations, which would play the role of phonons in conventional superconductivity. In this work, we explore spin-fluctuation-mediated pairing by studying a generalized gap equation with the pairing vertex replaced by a spin-fermion interaction. We use data from angle-resolved photoemission spectroscopy (ARPES) to construct the required Green's functions and spin susceptibilities. We find that spin fluctuations can yield superconducting critical temperatures of the order of 60–90 K, in agreement with experiments.

Over thirty years after their discovery, a coherent, microscopic description of copper-based High Temperature Superconductors (HTSCs) remains an outstanding problem in condensed matter physics. Because their electronic structure is primarily governed by the hybridization of copper (d) and oxygen (p) orbitals confined to two-dimensional sheets in their unit cell (the CuO_2 planes), these materials are commonly known as *cuprates*. Despite some important anomalies which we will address in this work, superconductivity in these materials can be understood within the theory put forth by Bardeen, Cooper, and Schrieffer (BCS) in 1957 [1] and extensions thereof. However, superconductivity is just one ordered phase, and the cuprates exhibit a host of other electronic phases which challenge our current understanding of solid state physics. The remarkable complexity of these compounds is evident from their generic phase diagram, where the various phases occupy

F. Restrepo, *Angle-Resolved Photoemission Spectroscopy Study of Spin Fluctuations in the Cuprate Superconductors*, Springer Theses,
https://doi.org/10.1007/978-3-031-10979-9_1

different regions of temperature and charge carrier concentration (see Fig. 1.1).[1] The parent (undoped) compounds are predicted by band theory to be metallic, since the antibonding band (crossing the Fermi energy) resulting from Cu-O hybridization is half-filled. However, due to strong on-site Coulomb repulsion, electron hopping is suppressed, leading to insulating behavior—the "Mott insulator" [2]. Adding charge carriers (or *doping*, for short) brings the system to a superconducting dome-shaped phase that persists over a wide doping range and has a maximum critical temperature (T_c) at an "optimal" doping. Beyond this doping level, electron correlations in the cuprates peter out and the system can be properly described as a Fermi liquid (that is, a normal metal) [3] at low temperatures.

However, below optimal doping and above the superconducting dome the system exhibits one of its most puzzling phases, the so-called *pseudogap* regime where, in addition to persisting strong electron correlations, there is a mysterious gap in the density of states centered at the Fermi energy. Other examples of ordered states observed in this region of the phase diagram include Charge Density Waves (CDW), Spin Density Waves (SDW), charge and spin "stripes" [4, 5], and even electronic analogs of liquid crystals like liquid, nematic, smectic, and "electronic solid" phases [6]. Finally, at even higher temperatures the cuprates are described as "strange metals", where macroscopic observables like the resistivity are governed not by quasi-particle interactions (as in conventional Fermi liquid theory) nor lattice imperfections, but are rather simply controlled by the temperature [7]. Figure 1.1 exhibits the most ubiquitous of these electronic phases. The dome-shaped lines represent the dependence of the superconducting transition temperature T_c on hole doping. The T^* line is the doping-dependent temperature above which the pseudogap vanishes. This work will be mainly focused on the superconducting and pseudogap phases.

Theoretical proposals to describe these phases are quite numerous, so the present review will be limited to two of the most successful classes of theories so far. The first class of theories emphasizes the strong Coulomb repulsion present in the Mott insulator and examines the effects of this interaction on the electronic structure as charge carriers are doped into the system. Besides being a Mott insulator, the undoped cuprate ground state is antiferromagnetic (AFM), with magnetic moments in the Copper ions interacting via a super-exchange interaction J (see Chap. 2). Anderson suggested that such antiferromagnet could be represented as a linear superposition of configurations of spin singlets, called the Resonating Valence Bond (RVB) state [9]. Anderson further s0howed that this state could be obtained by projecting out of the BCS ground state [1]

$$|\Phi\rangle = \prod_k u_k + v_k c^\dagger_{k,\uparrow} c^\dagger_{-k,\downarrow} |0\rangle \tag{1.1}$$

[1] Depending on whether the doped charge carriers are negatively or positively charged, one either speaks of electron- or hole-doped cuprates. In this work, we consider only the latter.

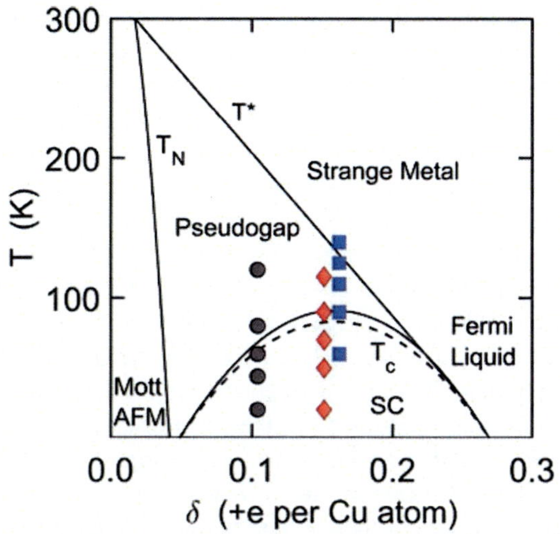

Fig. 1.1 Schematic phase diagram of $Bi_2Sr_2CaCu_2O_{8+\delta}$, the cuprate studied in this work. Although several other electronic phases have been reported, only those that are truly generic for all the cuprates are depicted. Symbols correspond to the points where photoemission data were collected and analyzed in this work. The solid and dashed T_c domes correspond to single crystal and thin film samples, respectively. The latter tend to exhibit lower T_c's and somewhat flattened domes [8]. (Adapted from Restrepo et al. Coupled electronic and magnetic excitations in the cuprates and their role in the superconducting transition. *Commun. Phys.* **5**, 45 (2022))

the factors corresponding to double occupancy at any given lattice site. By eliminating double occupancy, such a projection effectively suppresses electron (or hole) hopping between singly-occupied lattice sites and, hence, leads to insulating behavior. Thus, the Mott insulator can be described as a state resembling a conventional superconductor but with strong on-site repulsion built in. In this theory, superconductivity proper emerges when charge carriers are doped into the RVB state, whereupon the resonating spin singlets assume the role of conventional Cooper pairs.[2]

The RVB state has been the starting point of a myriad theoretical approaches describing the underdoped cuprates. These approaches have collectively succeeded in explaining a number of cuprate anomalies, notably the d-wave symmetry of the superconducting order and pseudogap phenomena (see Chap. 2). In these theories, the temperature scale T^* marks the onset of phase-incoherent Cooper pair formation. These charge-carrying singlets only become superconducting below T_c, where they achieve phase coherence. The general idea of "doping a Mott insulator" has also brought into sharp focus the connection between the superfluid density (essentially the concentration of charge carriers, which increases with doping) and

[2] This implementation of the RVB theory, based on the *Gutzwiller projection* [10] and the *t-J* model [11], is only one of numerous attempts to formally develop Anderson's original idea.

the superconducting phase stiffness, and its effect on the superconducting transition temperature. By virtue of the "uncertainty" relation between the phase fluctuations and number density, a lightly doped Mott insulator exhibits strong fluctuations which tend to depress T_c [12]. Another interesting prediction of doping a Mott insulator is the existence of instabilities toward charge and spin segregation into unidirectional "stripes", which have been detected in the Lanthanum-based cuprate and "nickelate" (where the transition metal Cu is substituted for Ni) families [4, 5, 13, 14], as well as in the Yttrium-based cuprate $YBa_2Cu_3O_{6+x}$ [15], although in this case stripe ordering has been only indirectly inferred from neutron scattering.

The second class of theories attempts to describe some of these phases as resulting from modifications of a Fermi liquid coming from AFM correlations. These approaches emphasize the fact that magnetism persists beyond the antiferromagnetic phase boundary (T_N line in Fig. 1.1) in the form of *spin fluctuations* (SFs). These excitations can be regarded as signatures of vestigial AFM correlations which extend over the phase diagram up to optimal doping and presumably up to the pseudogap temperature T^*. In these theories, the strong interaction between these bosonic modes and electrons—the *spin-fermion interaction*—is invoked to account for a variety of experimental observations. For example, in the superconducting state, two of the most controversial cuprate anomalies are the so-called spin resonance seen in Inelastic Neutron Scattering (INS) [16–18] and the peak-dip-hump structure of antinodal quasiparticles detected in Angle-resolved Photoemission Spectroscopy (ARPES) experiments [19].[3] Assuming the electron self-energy is solely determined by spin-fermion interactions, this quantity is proportional to the convolution of the electron and spin densities of states. The convolution of an electron density of states (with a superconducting gap Δ) and a spin density of states (peaked at the resonance energy Ω_0) yields a self-energy with a singularity at the energy

$$E_{\text{dip}} = \Delta + \Omega_0. \tag{1.2}$$

Strikingly, this is precisely the energy at which the "dip" in the ARPES spectra has been observed. Similar considerations account for dip features seen in SIS (superconductor-insulator-superconductor) tunneling experiments [20]. These experiments have shown that the above relation also holds at various doping levels, where the ARPES gaps are different. A singularity in the electron self-energy also gives rise to experimentally observed peak and dip features in the total density of states, the SIN (superconductor-insulator-normal metal) tunneling conductivity, and the optical and Raman responses, with energies dependent on the superconducting gap and the resonance energy [21]. The above experimental findings suggest that much of the cuprate fermiology can be accounted for by allowing electrons in these

[3] We do not wish at present to dwell on the physical origin of the spin resonance (although we shall re-visit this issue in Chap. 5), but simply point out the effects of its interaction with electronic excitations.

materials to be "dressed" (or *renormalized*) by SFs. Such an interplay between bosonic and fermionic excitations is especially significant if one recalls that similar mutual renormalizations between fermionic and bosonic degrees of freedom were seen in phonon-mediated superconductors, and their quantitative description based on the Eliashberg formalism [22] constituted an important validation of the BCS theory of superconductivity.

Pursuing this analogy with conventional superconductivity further, it is natural to inquire whether SFs can mediate superconductivity in the cuprates in the same way phonons do in conventional superconductors. The main question addressed in this work is, then, *is the spin-fermion interaction sufficiently strong to mediate a superconducting transition at experimentally observed temperatures (~90 K)?*

Similar to phonon-mediated superconductivity, determining the critical superconducting temperature in the cuprates involves solving an integral equation for a generalized order parameter $\Phi(\mathbf{k}, \omega_n, T)$—the Bethe-Salpeter equation (BSE) [23–26]:

$$\lambda(T)\Phi(T) = \hat{O}(T)\Phi(T). \tag{1.3}$$

A superconducting phase transition takes place at a temperature T_0 if a solution $\Phi(T_0)$ exists such that $\lambda(T_0) = 1$. The integral operator $\hat{O}(T)$ is defined at each temperature through

$$\begin{aligned} &\hat{O}(T)\Phi(\mathbf{k}, \omega_n, T) \\ &= -\frac{T}{N}\sum_{\mathbf{k}',\Omega_n} V(\mathbf{k}-\mathbf{k}', i\omega_n - i\Omega_n, T)G(\mathbf{k}', i\Omega_n, T)G(-\mathbf{k}', -i\Omega_n, T)\Phi(\mathbf{k}', \Omega_n, T), \end{aligned} \tag{1.4}$$

where the product of Green's functions is sometimes called the "pairing kernel" and V is the effective pairing interaction. Equations 1.3 and 1.4 are quite universal in that the specific mechanism leading to the emergence of superconductivity is reflected only in the form of V. In this work, we consider a proposed form of this pairing vertex where SFs mediate electron pairing [27]:

$$V = V_{\mathrm{SF}}(\mathbf{q}, i\omega_m, T) = \frac{3}{2}U_0(T)^2\chi(\mathbf{q}, i\omega_m, T), \tag{1.5}$$

where χ is the spin susceptibility and U_0 is an effective spin-fermion coupling.

Solution of the eigenvalue problem in Eqs. 1.3 and 1.4 requires the operator $\hat{O}$, which in turn depends on the Green's function, the coupling energy, and the spin susceptibility of the many-particle system in question. Purely theoretical approaches to this problem are limited by the computational cost of calculating the Green's function and the spin susceptibility for large systems of interacting particles, compounded with the fact that these functions also need to be known at different temperatures if the condition $\lambda(T) = 1$ is being sought. Alternatively, some of these quantities could be obtained from experiments. In particular, probes which couple to magnetic excitations—like Nuclear Magnetic Resonance (NMR) and INS—can be used to extract the spin susceptibility, while probes that couple to electronic excitations—like ARPES—may be used to obtain the Green's function. (As shown in Chap. 5, in this work we obtained *both* the electronic and magnetic excitation spectra from ARPES data.)

Studies where experimental data are used to solve the BSE are sparse and have yielded conflicting results for the strength of the SF-induced pairing interaction. Monthoux and Pines [28] considered an NMR-derived form of the spin susceptibility for the cuprate $YBa_2Cu_3O_7$ [29] as an input to Eq. 1.3. By solving a set of self-consistent Eliashberg equations for the Green's function, the self-energy, and the coupling constant, they concluded that spin fluctuations could mediate a superconducting transition at a temperature around 90 K, with a coupling constant of order 1 eV. In a similar vein, Dahm et al. [30] also used the spin susceptibility as input to the BSE, but using the more recent, high-resolution data available from INS experiments on the cuprate $YBa_2Cu_3O_{6.6}$. As in Ref. [28], the Green's function was not known and had to be calculated self-consistently along with the self-energy. The authors found transition temperatures of order 150 K and concluded, like Monthoux et al., that spin fluctuations could account for high temperature superconductivity. For another widely studied cuprate, $Bi_2Sr_2CaCu_2O_{8+\delta}$, Mishra et al. [25] obtained the Green's functions directly from ARPES data taken at 140 K and from them computed the spin susceptibility via the Random Phase Approximation (RPA). Then, they solved Eq. 1.3 and found that $\lambda << 1$ at all temperatures, implying that V_{SF} was not strong enough to give a high T_c superconducting transition in the cuprates, in stark contrast to the findings of Refs. [28, 30]. In a subsequent theoretical study, Maier et al. [26] ascribed the failure of Mishra et al. to find any superconducting transition to the fact that the experimental ARPES data were taken at a temperature of 140 K, well above the critical temperature of $T_c = 90$ K.

In Refs. [28, 30] the single-particle Green's function was indirectly inferred from the spin susceptibility, which is a two-particle correlation function. Mishra et al. followed the inverse procedure of piecing together the two-particle response from the single-particle Green's function probed in ARPES. On the other hand, these three works considered data sets limited to one temperature (well above T_c) and reached opposite conclusions regarding the strength of the SF-mediated pairing interaction. As noted in the work of Maier et al., the problem would be best elucidated if data collected at T_c were employed in the solution of the BSE. Taken together, these studies have suggested that, in order to determine whether the spin-fermion interaction can be the source of the unconventional superconducting state in the cuprate superconductors, three requirements need to be satisfied: (1) experimental ARPES data need to be taken above, but close to, T_c, (2) the effective superconducting pairing vertex needs to be computed from these ARPES data, and (3) the solution of the gap equation, using the ARPES data and the derived pairing vertex as input, should reproduce the transition temperature of the material.

Here, we used ARPES data from three underdoped and optimally doped $Bi_2Sr_2CaCu_2O_{8+\delta}$ (Bi2212) samples, taken at temperatures below and above T_c

(as indicated by the symbols in Fig. 1.1), to construct the spin-fluctuation-mediated pairing vertex and solve the Bethe-Salpeter gap equation. This yielded a superconducting order parameter with $d_{x^2-y^2}$-wave symmetry, and a critical temperature for all samples close to the ones observed experimentally. Although our main goal was to solve the BSE near T_c and verify whether the SF-mediated pairing interaction could support high-T_c superconductivity, other significant findings sprang from this effort, as summarized below:

- The spin susceptibilities constructed from ARPES data within the RPA show satisfying agreement with those measured in INS. To our knowledge, there are two works where such a construction was attempted [25, 31], but the emphasis was placed primarily on reproducing well-known features in the momentum structure of this function. Moreover, both of these studies focused on a single point in the phase diagram (that is, one doping, one temperature). We extended the methodology therein to show that it can semi-quantitatively account for the temperature and doping dependence of spin excitations observed in INS. In particular, we focused mainly on capturing the temperature and doping dependence of the most salient feature of the cuprate spin response, the "spin resonance" (and its broadened precursor in the pseudogap state), while still reproducing the universal "upper branch" of the spin excitation spectrum (see Chaps. 2 and 5). More quantitatively, our calculated spin susceptibilities show spectral weight distributions below T_c in reasonable agreement with those inferred from INS data from Yttrium-based cuprates.
- The "symmetrization" procedure [32] commonly employed to extend ARPES data to cover the unoccupied states leads to artifacts in the spin susceptibility. These artifacts show up as anomalously strong spectral weight in the spin susceptibility near $\mathbf{q} = (0, 0)$ and are particularly deleterious to SF-mediated superconductivity, where the pairing interaction is assumed to be peaked around $\mathbf{Q} = (\pi, \pi)$. We believe this observation may be of use in calculations where two-particle correlation functions (like the spin susceptibility, the optical conductivity, or the Raman scattering cross-section) are obtained from the single-particle spectra measured in ARPES.
- The generalized order parameter $\Phi(\mathbf{k}, \omega_n)$, obtained in Matsubara frequencies, has d-wave symmetry, is even in frequency, and decays with a characteristic energy scale of $\sim$250 meV. Following numerical solutions of the widely studied Hubbard model [33], where this decay is of order $2J$ (J being the exchange interaction) [23], we extract values of the exchange coupling of order 125 meV, in agreement with values obtained from INS on other cuprates. This energy decay implies a retardation in time of the pairing interaction $(2J)^{-1} \sim 10$ fs, which has been observed in ultrafast optical spectroscopy measurements [34].
- The main result of this study is that, near the experimental T_c's, the ARPES-derived eigenvalues approach unity for the three samples considered. That is, the spin-fermion interaction can mediate a high-temperature superconducting transition. Extending our calculations to the pseudogap region above T_c, we see that the eigenvalue decreases with increasing temperature, a behavior which

has been predicted in numerical studies of the two-dimensional Hubbard model for the cuprates [26, 35, 36]. Collectively, our results suggest that the RPA and the spin fluctuation framework may provide a consistent description of Bi2212 connecting the electronic (ARPES) and magnetic (INS) responses and yielding a high-temperature superconducting transition to a state with d-wave symmetry.

In order to better understand these results, the first half of this thesis will cover some necessary background material. Chapter 2 contains a brief introduction to conventional superconductivity, the gap equation, and a description of the spin-fluctuation-induced pairing mechanism. This chapter also covers the basic chemistry and electronic structure of the compound studied in this work, Bi2212, and ends with a survey of experimental results indicating the presence of spin fluctuations in the cuprates and identifying their possible role in mediating a superconducting transition in these materials. Chapter 3 contains a description of the experimental technique ARPES and an introduction to the important quantity it probes: the *spectral function*. Chapter 4 covers some experimental aspects of this work, with an emphasis on the author's contributions to the design and construction of the measurement apparatus. Also included in this chapter are results which are interesting from a purely experimental standpoint, having to do with extending the lifetime of Bi2212 samples in ultrahigh vacuum.

In Chap. 5, we present the main findings of this work. There, it is shown how photoemission data were used to construct the pairing interaction, solve the Bethe-Salpeter gap equation, and obtain estimates for the superconducting critical temperature. Finally, conclusions are presented in Chap. 6. Some ancillary material on the anomalous Green's function, Fourier transforms in Matsubara representation, and the solution method for the BSE is presented in the Appendices.

References

1. J. Bardeen, L.N. Cooper, J.R. Schrieffer, Theory of superconductivity. Phys. Rev. **108**(5), 1175–1204 (1957)
2. N.F. Mott, The transition to the metallic state. Philos. Mag. J. Theoret. Exp. Appl. Phys. **6**(62), 287–309 (1961)
3. E.M. Lifshitz, L.P. Pitaevski, *Statistical Physics, Part 2. Course of Theoretical Physics. Volume 9* (Butterworth-Heinemann, Oxford, 1980)
4. J.M. Tranquada, B.J. Sternlieb, J.D. Axe, Y. Nakamura, S. Uchida, Evidence for stripe correlations of spins and holes in copper oxide superconductors. Nature **375**, 561 (1995)
5. P. Abbamonte, A. Rusydi, S. Smadici, G.D. Gu, G.A. Sawatzky, D.L. Feng, Spatially modulated 'Mottness' in $La_{2-x}Ba_xCuO_4$. Nat. Phys. **1**, 155 (2005)
6. S.A. Kivelson, E. Fradkin, V.J. Emery, Electronic liquid-crystal phases of a doped Mott insulator. Nature **393**, 550 (1998)
7. P. Giraldo-Gallo, J.A. Galvis, Z. Stegen, K.A. Modic, F.F. Balakirev, J.B. Betts, X. Lian, C. Moir, S.C. Riggs, J. Wu, A.T. Bollinger, X. He, I. Božović, B.J. Ramshaw, R.D. McDonald, G.S. Boebinger, A. Shekhter, Scale-invariant magnetoresistance in a cuprate superconductor. Science **361**(6401), 479–481 (2018)

8. Z. Konstantinovic, Z.Z. Li, H. Raffy, Normal state transport properties of single- and double-layered $Bi_2Sr_2Ca_{n-1}Cu_nO_y$ thin films and the pseudogap effect. Physica C Superconductivity **341–348**, 859–862 (2000)
9. P.W. Anderson, The resonating valence bond state in La_2CuO_4 and superconductivity. Science **235**(4793), 1196–1198 (1987)
10. M.C. Gutzwiller, Effect of correlation on the ferromagnetism of transition metals. Phys. Rev. Lett. **10**(5), 159–162 (1963)
11. P.W. Anderson, P.A. Lee, M. Randeria, T.M. Rice, N. Trivedi, F.C. Zhang, The physics behind high-temperature superconducting cuprates: the 'plain vanilla' version of RVB. J. Phys. Condensed Matter **16**(24), R755 (2004)
12. V.J. Emery, S.A. Kivelson, Importance of phase fluctuations in superconductors with small superfluid density. Nature **374**(6521), 434–437 (1995)
13. A. Vigliante, T. Niemöller, N. Ichikawa, T. Frello, J. Madsen, P. Wochner, S. Uchida, N.H. Andersen, J.M. Tranquada, D. Gibbs, J.R. Schneider, Hard–X-ray diffraction study of charge stripe order in $La_{1.48}Nd_{0.4}Sr_{0.12}CuO_4$. Europhys. Lett. **41**(6), 629–634 (1998)
14. S.M. Hayden, G.H. Lander, J. Zarestky, P.J. Brown, C. Stassis, P. Metcalf, J.M. Honig, Incommensurate magnetic correlations in $La_{1.8}Sr_{0.2}NiO_4$. Phys. Rev. Lett. **68**(7), 1061–1064 (1992)
15. H.A. Mook, P. Dai, S.M. Hayden, G. Aeppli, T.G. Perring, F. Doğan, Spin fluctuations in $YBa_2Cu_3O_{6.6}$. Nature **395**(6702), 580–582 (1998)
16. J. Rossat-Mignod, L.P. Regnault, C. Vettier, P. Bourges, P. Burlet, J. Bossy, J.Y. Henry, G. Lapertot, Neutron scattering study of the $YBa_2Cu_3O_{6+x}$ system. Physica C Superconductivity **185–189**, 86–92 (1991)
17. H.F. Fong, P. Bourges, Y. Sidis, L.P. Regnault, A. Ivanov, G.D. Gu, N. Koshizuka, B. Keimer, Neutron scattering from magnetic excitations in $Bi_2Sr_2CaCu_2O_{8+\delta}$. Nature **398**(6728), 588–591 (1999)
18. H.F. Fong, B. Keimer, D.L. Milius, I.A. Aksay, Superconductivity-induced anomalies in the spin excitation spectra of underdoped $YBa_2Cu_3O_{6+x}$. Phys. Rev. Lett. **78**(4), 713–716 (1997)
19. A. Kaminski, M. Randeria, J.C. Campuzano, M.R. Norman, H. Fretwell, J. Mesot, T. Sato, T. Takahashi, K. Kadowaki, Renormalization of the spectral line shape and dispersion below T_c in $Bi_2Sr_2CaCu_2O_{8+\delta}$. Phys. Rev. Lett. **86**(6), 1070–1073 (2001)
20. J.F. Zasadzinski, L. Ozyuzer, N. Miyakawa, K.E. Gray, D.G. Hinks, C. Kendziora, Correlation of tunneling spectra in $Bi_2Sr_2CaCu_2O_{8+\delta}$ with the resonance spin excitation. Phys. Rev. Lett. **87**(6), 067005 (2001)
21. A.V. Chubukov, D. Pines, J. Schmalian, A spin fluctuation model for d-wave superconductivity, in *Superconductivity: Conventional and Unconventional Superconductors*, ed. by K.H. Bennemann, J.B. Ketterson (Springer, Berlin, 2008), pp. 1349–1413
22. W.L. McMillan, J.M. Rowell, Lead phonon spectrum calculated from superconducting density of states. Phys. Rev. Lett. **14**(4), 108–112 (1965)
23. T.A. Maier, M. Jarrell, D.J. Scalapino, Pairing interaction in the two-dimensional Hubbard model studied with a dynamic cluster quantum Monte Carlo approximation. Phys. Rev. B **74**(9), 094513 (2006)
24. G.P. Malik, *Superconductivity: A New Approach Based on the Bethe-salpeter Equation in the Mean-field Approximation*. Series on Directions in Condensed Matter Physics (World Scientific Publishing Company, 2016)
25. V. Mishra, U. Chatterjee, J.C. Campuzano, M.R. Norman, Effect of the pseudogap on the transition temperature in the cuprates and implications for its origin. Nat. Phys. **10**, 357 (2014)
26. T.A. Maier, P. Staar, V. Mishra, U. Chatterjee, J.C. Campuzano, D.J. Scalapino, Pairing in a dry Fermi sea. Nat. Commun. **7**, 11875 (2016)
27. J.R. Schrieffer, X.G. Wen, S.C. Zhang, Dynamic spin fluctuations and the bag mechanism of high-T_c superconductivity. Phys. Rev. B **39**(16), 11663–11679 (1989)
28. P. Monthoux, D. Pines, Spin-fluctuation-induced superconductivity in the copper oxides: A strong coupling calculation. Phys. Rev. Lett. **69**(6), 961–964 (1992)

29. A.J. Millis, H. Monien, D. Pines, Phenomenological model of nuclear relaxation in the normal state of $YBa_2Cu_3O_7$. Phys. Rev. B **42**(1), 167–178 (1990)
30. T. Dahm, V. Hinkov, S.V. Borisenko, A.A. Kordyuk, V.B. Zabolotnyy, J. Fink, B. Büchner, D.J. Scalapino, W. Hanke, B. Keimer, Strength of the spin-fluctuation-mediated pairing interaction in a high-temperature superconductor. Nat. Phys. **5**(3), 217–221 (2009)
31. U. Chatterjee, D.K. Morr, M.R. Norman, M. Randeria, A. Kanigel, M. Shi, E. Rossi, A. Kaminski, H.M. Fretwell, S. Rosenkranz, K. Kadowaki, J.C. Campuzano, Dynamic spin-response function of the high-temperature $Bi_2Sr_2CaCu_2O_{8+\delta}$ superconductor from angle-resolved photoemission spectra. Phys. Rev. B **75**(17), 172504 (2007)
32. M.R. Norman, H. Ding, M. Randeria, J.C. Campuzano, T. Yokoya, T. Takeuchi, T. Takahashi, T. Mochiku, K. Kadowaki, P. Guptasarma, D.G. Hinks, Destruction of the Fermi surface in underdoped high-T_c superconductors. Nature **392**, 157 (1998)
33. R.A. Jishi, *Feynman Diagram Techniques in Condensed Matter Physics*, 1st edn. (Cambridge University Press, New York, 2013)
34. S. Dal Conte, L. Vidmar, D. Golež, M. Mierzejewski, G. Soavi, S. Peli, F. Banfi, G. Ferrini, R. Comin, B.M. Ludbrook, L. Chauviere, N.D. Zhigadlo, H. Eisaki, M. Greven, S. Lupi, A. Damascelli, D. Brida, M. Capone, J. Bonča, G. Cerullo, C. Giannetti, Snapshots of the retarded interaction of charge carriers with ultrafast fluctuations in cuprates. Nat. Phys. **11**(5), 421–426 (2015)
35. N. Bulut, D.J. Scalapino, S.R. White, Bethe-Salpeter eigenvalues and amplitudes for the half-filled two-dimensional Hubbard model. Phys. Rev. B **47**(21), 14599–14602 (1993)
36. T.A. Maier, M.S. Jarrell, D.J. Scalapino, Structure of the pairing interaction in the two-dimensional Hubbard model. Phys. Rev. Lett. **96**(4), 047005 (2006)

Chapter 2
Superconductivity and the Cuprates

Abstract The phenomena of infinite conductivity and magnetic flux expulsion in elementary metals at low temperatures were not fully understood until the microscopic theory of Bardeen, Cooper, and Schrieffer (BCS) was put forth in 1957. Despite the enormous success of this theory and its subsequent generalizations to cases where phonon dynamics could no longer be ignored, superconductivity in the more recently discovered cuprate superconductors seems to elude a simple understanding based on lattice-mediated pairing. Of the various proposals to rationalize the anomalous behavior of these materials, extensions of the BCS theory where electrons are bound not by phonons, but by magnetic excitations, are among the most developed. The purpose of this chapter is threefold: (1) it reviews the basic ideas behind the BCS theory in Green's function formalism, (2) it presents a review of the electronic structure of the cuprates, emphasizing the points of departure from the BCS theory, and (3) it presents a brief survey of the evidence for magnetic excitations in these materials and their possible role in mediating Cooper pairing.

The first section of this chapter summarizes the BCS theory of conventional superconductivity. The approach followed is that of Gorkov (as summarized in reference [1]), for three reasons. First, it directly introduces Matsubara Green's functions, which will be used extensively in Chap. 5. Second, it provides a quick pathway to the important *gap equation*, which plays a central role in this work. Finally, this formalism is more readily generalized to more sophisticated treatments, like that of Eliashberg and the spin-fluctuation-induced pairing mechanism. Section 2.2 describes the basic chemistry and electronic structure of the cuprates (with an emphasis on $Bi_2Sr_2CaCu_2O_{8+\delta}$, the compound studied in this work). Section 2.3 presents an overview at the cuprate phase diagram, with an emphasis on the superconducting and pseudogap phases. Finally, Sect. 2.4 reviews experimental evidence for an anisotropic (in momentum space) pairing interaction and its consequences for superconductivity in the cuprates.

F. Restrepo, *Angle-Resolved Photoemission Spectroscopy Study of Spin Fluctuations in the Cuprate Superconductors*, Springer Theses,
https://doi.org/10.1007/978-3-031-10979-9_2

2.1 BCS Superconductivity

As its name suggests, superconductivity is a phenomenon in which a material exhibits an undetectably small resistivity to electrical current below some *critical temperature* T_c [2]. Aside from this remarkable property, soon after their discovery, these materials were quickly found to exhibit even more unusual behavior when subjected to magnetic fields. An applied magnetic field above T_c (in the resistive state) should penetrate the material, due to the latter's inability to host currents opposing the change in magnetic flux. Cooling the sample below T_c (superconducting state) should leave this configuration unaltered because, even if charges are now free to move around, there is no change in magnetic flux to compete against. However, the famous experiments of Meissner and Oschenfeld [3] showed the opposite behavior: a transition into the superconducting state resulted in the expulsion of the magnetic field that was initially allowed into the material. This "Meissner" effect was a clear departure from classical electrodynamics and suggested the presence of subtle quantum-mechanical effects driving an electronic phase transition into a hitherto unknown state of matter.

The development of a theory of superconductivity was a long and tortuous process, primarily because in the years after its discovery theorists relied heavily on perturbative techniques. While tremendously successful in other problems of condensed matter physics, these approaches ultimately proved incapable of explaining the new effect. To shorten what would otherwise be an unnecessarily long story, we have selected three milestones in the development of the theory of superconductivity as most directly conducive to its final formulation:

- The observation that the ionic lattice played an important role in the phenomenon, as evidenced by the relation $T_c \sim M^{-1/2}$. Roughly speaking, this relation—known as the *isotope effect*—suggests that the transition temperature scales with a characteristic phonon energy.
- Cooper's observation that the Pauli principle imposed phase space restrictions on electrons with opposite spin and momenta near the Fermi surface that could lead them to *pair*, and that such pairing could lower the energy of the system.
- The realization that the phenomena of zero resistivity and perfect diamagnetism (along with anomalies in the heat capacity, tunneling spectroscopy, and other experiments available at the time) could be explained by the existence of an energy gap in the single-particle excitation spectrum of these materials.

The obvious issue with Cooper's idea was Coulomb repulsion, which acts to suppress pairing. Motivated by the first observation, Bardeen and Pines [4] showed how a phonon-mediated, retarded interaction could exceed the Coulomb contribution, leading to an effective attractive force. Their argument is commonly visualized as follows: an electron attracts its neighboring ions to create a transient, local excess positive charge, which then may attract another electron. The main point is that the interaction is *retarded*; that is, the second electron is drawn towards the spatial coordinate of the first after the latter has perturbed its ionic environment and moved elsewhere. In the BCS picture, phonons were regarded as "mediators" of superconductivity, and the "pairing strength" was identified as being dependent on the phonon spectrum.

Having set the notion of electron pairing on a firm basis, Bardeen, Cooper, and Schrieffer considered the Hamiltonian [5]

$$H = \sum_{k,\sigma} \epsilon_k c^\dagger_{k\sigma} c_{k\sigma} + \sum_{k,k'} V_{kk'} c^\dagger_{-k'\downarrow} c^\dagger_{k'\uparrow} c_{k\uparrow} c_{-k\downarrow}. \tag{2.1}$$

The operator $c^\dagger_{k\sigma}$ creates an electron in the Bloch state with quantum numbers $k\sigma$ (k representing momentum and σ the spin) and energy ϵ_k (single-particle energies are measured with respect to the chemical potential, μ). The first term is diagonal and corresponds to the kinetic energy. The second term represents electron interactions, which in this simplified Hamiltonian are only of the Cooper type. The restriction to pairing between electrons of opposite momenta owes to the fact that zero center-of-mass momentum minimizes the energy at equilibrium (i.e. no current). $V_{kk'}$ is the phonon-mediated electron interaction, assumed to be negative (attractive).

To derive the excitation spectrum of the BCS Hamiltonian, Gorkov's approach makes use of *imaginary time* Green's functions, where the time variable is replaced by an *imaginary time*: $t \to -i\tau$. The single-particle destruction operator has a τ-dependence of the Heisenberg form $c_{k\sigma}(\tau) = e^{H\tau} c_{k\sigma} e^{-H\tau}$, where H is the BCS Hamiltonian (Eq. 2.1). Then, the Green's function is defined by

$$G(k,\tau) = -\langle T_\tau c_{k\uparrow}(\tau) c^\dagger_{k\uparrow}(0)\rangle, \tag{2.2}$$

where T_τ is the τ-ordering operator and the brackets denote a thermodynamic average in the grand canonical ensemble. This Green's function is defined in the interval $(-\beta, \beta)$ (where $\beta = 1/k_B T$) and is anti-periodic under $\tau \to \tau + \beta$, from which fact it can be resolved into Fourier components:

$$G(k,\tau) = \frac{1}{\beta} \sum_{\omega_n} e^{-i\omega_n \tau} G(k, \omega_n), \tag{2.3}$$

where only the Fermionic *Matsubara frequencies*, $\omega_n = (2n+1)\pi/\beta$, are included in the sum. The above equation can be inverted to yield the Fourier coefficients

$$G(k, i\omega_n) = \frac{1}{2} \int_{-\beta}^{\beta} d\tau e^{i\omega_n \tau} G(k,\tau). \tag{2.4}$$

Having defined the imaginary time Green's function and established a few of its properties, Gorkov's argument begins by taking a τ-derivative of this quantity:

$$\left(-\frac{d}{d\tau} - \epsilon_k\right) G(k,\tau) = \delta(\tau) + \sum_{k'} V_{kk'} \langle T_\tau c^\dagger_{-k\downarrow}(\tau) c_{k'\uparrow}(\tau) c_{-k'\downarrow}(\tau) c^\dagger_{k\uparrow}(0)\rangle. \tag{2.5}$$

To solve this equation of motion for $G(k,\tau)$, the thermal average is evaluated using Wick's theorem [6], where four-operator averages like the last term in the above equation are decomposed into 2-operator averages according to the prescription:

$$\langle T_\tau c^\dagger_{-k\downarrow}(\tau)c_{k'\uparrow}(\tau)c_{-k'\downarrow}(\tau)c^\dagger_{k\uparrow}(0)\rangle \to \langle T_\tau c_{k'\uparrow}(\tau)c_{-k'\downarrow}(\tau)\rangle\langle T_\tau c^\dagger_{-k\downarrow}(\tau)c^\dagger_{k\uparrow}(0)\rangle. \tag{2.6}$$

There are more terms in the above decomposition; however, they correspond to the usual combinations of the form $c^\dagger_{k_1\sigma}(\tau)c_{k_2\beta}(\tau)$ typical of the Hartree Fock (HF) theory. These terms should appear in both the normal and superconducting states, so they will cancel out when computing, for example, the condensation energy. Therefore, only the "anomalous" terms, with two creation or destruction operators, are considered. These lead to the definition of the *anomalous Green's function* and its complex conjugate:

$$F(k,\tau) = -\langle T_\tau c_{k\uparrow}(\tau)c_{-k\downarrow}(0)\rangle \tag{2.7}$$

$$F^*(k,\tau) = -\langle T_\tau c^\dagger_{-k\downarrow}(\tau)c^\dagger_{k\uparrow}(0)\rangle. \tag{2.8}$$

In conventional HF theory, such terms vanish by conservation of particle number, a restriction no longer enforced in the BCS theory. F^* and F are the amplitudes for creating or destroying two electrons with opposite spin and momenta; that is, the *Cooper pairs* mentioned earlier. Inserting these definitions into Eq. 2.5 and taking the τ-derivative of $F^*(k,\tau)$ yields the following coupled, nonlinear differential equations:

$$\left(-\frac{d}{d\tau}-\epsilon_k\right)G(k,\tau) = \delta(\tau) + \sum_{k'} V_{kk'}F(k',0)F^*(k,\tau) \tag{2.9}$$

$$\left(-\frac{d}{d\tau}+\epsilon_k\right)F^*(k,\tau) = \sum_{k'} V_{kk'}F^*(k',0)G(k,\tau). \tag{2.10}$$

Defining the *gap function*

$$\Delta_k = -\sum_{k'} V_{kk'}F(k',0) \tag{2.11}$$

and using the Fourier representation in Eq. 2.3 (with a similar one for $F^*(k,\tau)$) leads to the matrix equation

$$\begin{pmatrix} i\omega_n-\epsilon_k & \Delta_k \\ \Delta_k^* & i\omega_n+\epsilon_k \end{pmatrix}\begin{pmatrix} G(k,i\omega_n) \\ F^*(k,i\omega_n) \end{pmatrix} = \begin{pmatrix} 1 \\ 0 \end{pmatrix}, \tag{2.12}$$

which can be inverted to yield the desired Green's functions:

$$G(k,\omega_n) = \frac{-i\omega_n - \epsilon_k}{\omega_n^2 + E_k^2} = \frac{1}{i\omega_n - E_k} \tag{2.13}$$

$$F^*(k,\omega_n) = \frac{\Delta_k^*}{\omega_n^2 + E_k^2}, \tag{2.14}$$

where both propagators have poles at the quasi-particle energy $E_k = \sqrt{\epsilon_k^2 + \Delta_k^2}$. Physically, the gap function Δ_k quantifies the energy required to break a Cooper pair. The problem is completely solved once the gap function is calculated. Since this function itself depends on F^*, the problem calls for a *self-consistent* solution. Combining Eqs. 2.11 and 2.14 leads to the celebrated *gap equation*

$$\Delta_k = -\frac{1}{\beta}\sum_{k'}\sum_{\omega_n} \frac{V_{kk'}\Delta_{k'}}{\omega_n^2 + E_{k'}^2}. \tag{2.15}$$

If $\Delta_k = \Delta$ (momentum-independent), solution of Eq. 2.15 yields a gap function which decays monotonically with temperature and approaches zero at a critical temperature T_c in the mean-field fashion $\Delta \sim \sqrt{1 - T/T_c}$ [5].

Consider the behavior of this equation in the vicinity of the critical temperature, where $\Delta_k \sim 0$. Then, $E_k \sim \epsilon_k$ and we may write

$$\Delta_k = -\frac{1}{\beta}\sum_{k'}\sum_{\omega_n} V_{kk'}G(k',\omega_n)G(-k',-\omega_n)\Delta_{k'}, \tag{2.16}$$

where the Green's functions are those of the normal state. Note the close resemblance between this equation and the Bethe-Salpeter equation mentioned in the introduction. Thus, we see how Gorkov's formulation leads to a gap equation similar to the BSE considered in this work (Eqs. 1.3 and 1.4) while leaving the pairing interaction undetermined, so it can be chosen to be of the spin-fluctuation type. For now, though, we continue with the BCS solution of this equation.

To calculate T_c, we invoke the BCS approximation $V_{kk'} = -V$ for $|\epsilon_k| \le \omega_D$ (where ω_D is the Debye frequency, which in this case provides the relevant energy scale for phonon-mediated coupling) and assume $\Delta_k = \Delta$ in the same energy range, so that Eq. 2.16 simplifies to

$$\frac{1}{N(0)V} = \int_0^{\hbar\omega_D} \frac{d\epsilon}{\epsilon}\tanh\frac{\epsilon\beta_c}{2}, \tag{2.17}$$

where the density of states at the Fermi energy, $N(0)$ (assumed constant in the range of integration), was introduced to convert the momentum sum into an energy integral.[1] For weak coupling, $k_B T_c << \omega_D$ and we obtain for the critical temperature:

$$\beta_c^{-1} = k_B T_c = 1.13\omega_D e^{-1/N(0)V}. \tag{2.18}$$

This is a manifestly non-perturbative result, as the coupling constant appears inside an exponential function.

A few remarks are in order before concluding this section:

- The phenomenon of superconductivity itself is explained by the energy gap, which depletes the number of single-particle states where electrons can be scattered into, thus suppressing resistivity.[2]
- A gap in the energy spectrum also leads naturally to the expressions for the current inside a superconductor in the presence of a magnetic field originally derived by London [8], from which the Meissner effect immediately follows.
- Thermodynamic quantities like the condensation energy and the discontinuity in specific heat may be derived from these results. Both these quantities are proportional to the gap Δ [6].
- The pairing gap need not be of the simple s-wave ($\Delta_k = \Delta$) form assumed above. In fact, the total antisymmetry requirement for the pair wavefunctions allows $s = 0$ (singlet) pairing as well as $s = 1$ (triplet) pairing. In a two-dimensional square lattice (appropriate for the cuprates, as discussed in the next section), the $s = 0$ subset allows for further decomposition into s-wave (no momentum dependence), extended s-wave ($\cos k_x + \cos k_y$), and *d-wave* pairing ($\cos k_x - \cos k_y$), which changes sign in crossing the Brillouin Zone (BZ) diagonals. The latter pairing symmetry is realized in the cuprates and arises naturally in theories where pairing is due to spin fluctuations, as discussed in Sect. 2.4.

2.2 Chemistry and Electronic Structure of the Undoped Cuprates

Figure 2.1a shows a unit cell in the crystal structure of Bi2212. The unit cell is approximately tetragonal, the base having sides $a = b = 5.4$ Å, whereas the height is $c = 30.7$ Å. This cell can be viewed as two blocks of stacked layers in the order BiO, SrO, CuO_2, Ca, CuO_2, SrO, and BiO, the two blocks being offset by a vector $(a/2, 0, c/2)$. The weak van der Waals coupling between BiO layers (which are about 3 Å apart [9]) provides a natural cleavage plane for this material, making it

[1] The sum over ω_n can be computed via the identity $\sum_n \frac{1}{(2n+1)^2\pi^2+a^2} = \frac{1}{2a}\tanh\frac{a}{2}$.

[2] It must be noted that a gap in the density of states is not a *necessary* condition for superconductivity. It is rather the existence of a Cooper pair condensate which guarantees infinite conductivity [7].

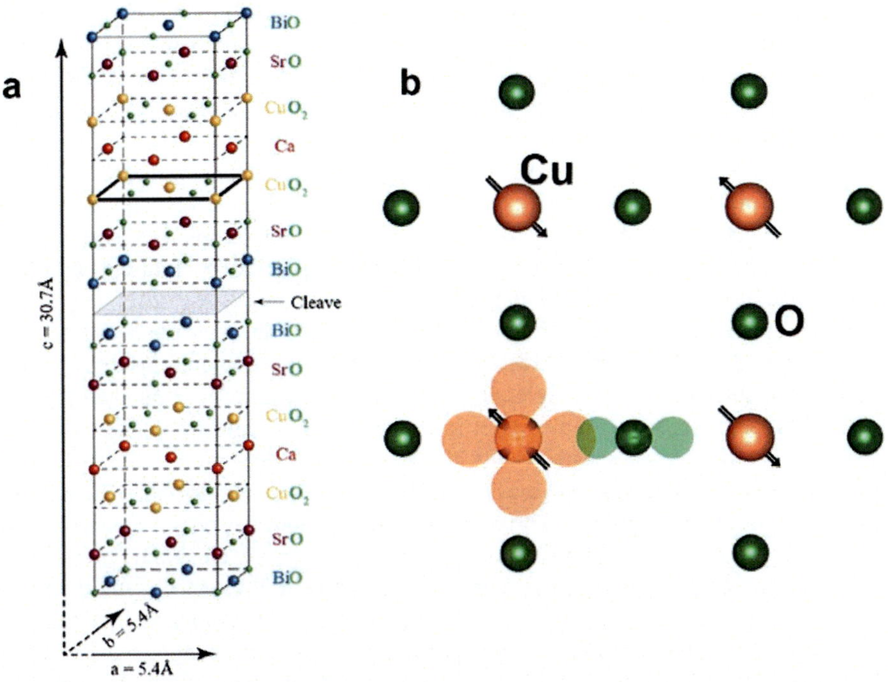

Fig. 2.1 (**a**) Crystal structure of the High-T_c superconductor Bi2212, showing the cleavage plane and a CuO_2 plane (thick lines). (Aside from the highlighted CuO_2 plane and the cleaving plane, this figure is a reprint from Wikipedia. Author: James Slezak. License: https://commons.wikimedia.org/wiki/File:Bi2212_Unit_Cell.png) (**b**) CuO_2 plane showing AFM spin alignment and schematic Cu *d* and O *p* orbitals

fairly easy to cleave mechanically (as described in Sect. 4.5). As a result, the layer immediately exposed is BiO, which is roughly 12 Å away from the closest CuO_2 layer below. Each copper atom in these CuO_2 layers is surrounded by four in-plane oxygen atoms and sits about 2 Å [9] directly below (above) an *apical* oxygen atom, belonging to the SrO plane above (below). The structure is known to possess a superlattice distortion where the atoms are displaced by up to 0.4 Å away from their ideal Bravais Lattice positions every 26 Å in the *a*-direction [10]. Even though there are indications that such structural modifications may lead to important effects on the electronic structure of these materials [10–12], this work shall focus mainly on the CuO_2 planes, where the key to the Cooper pairing mechanism is believed to reside [13]. The spacer layers between CuO_2 act as charge reservoirs for the CuO_2 planes.

In the CuO_2 planes, the passage to electronic structure is facilitated by molecular orbital considerations. An isolated Cu atom has electronic configuration $3d^{10}4s^1$, whereas an isolated O atom has $2s^22p^4$. A convenient starting point for the description of the Bi2212 crystal is by treating it as *ionic* solid, where electrons are localized around ion cores, but not necessarily their *parent* ions [14]. As a result, one can assign formal valences to each atom in the unit cell. In particular, copper has a +2 charge (Cu^{+2}), leaving it in a $3d^9$ configuration and oxygen has a −2 charge (O^{-2}), leaving it at $2s^2 2p^6$. As it turns out, the atoms in the spacer layers tend to keep their formal valences even as holes are doped into the system. However, in the crystalline environment, the energy of the copper d-orbitals is reduced enough to nearly coincide with that of the oxygen p-orbitals. This near-degeneracy leads to a *covalent* (more metal-like) bonding of copper and oxygen, whose valences can now vary with the degree of doping [9]. In the crystal, the anisotropic electrostatic fields created by other ions lead to a crystal field splitting of the fivefold degenerate Cu d-orbitals into two groups: the doubly degenerate $e_g = \{d_{x^2-y^2}, d_{3z^2-r^2}\}$, and the triply degenerate $t_{2g} = \{d_{xy}, d_{xz}, d_{yz}\}$. The resulting configuration is still unstable and, by the Jahn-Teller effect [14], the degeneracy of these two groups may be lifted further, resulting in three singlets $d_{x^2-y^2}$, $d_{3z^2-r^2}$, d_{yz}, and a doublet $\{d_{xz}, d_{yz}\}$. The oxygen p-orbitals are also split in the crystal field into three singlets $\pi_{\parallel}$, $\pi_{\perp}$ and σ. Of these three orbitals, only the in-plane σ point along the Cu-O bonds and experience a substantial overlap with the Cu $d_{x^2-y^2}$ orbitals. The different configurations of σ orbitals hybridize with the Cu $d_{x^2-y^2}$ orbitals leading to broad bonding (σ) and anti-bonding (σ^*) bands, as well as weakly dispersing non-bonding bands. The resulting antibonding band is half-filled and thus, from band theory, the system is expected to be metallic. However, due to a phenomenon first described by Mott [15], electron hopping is suppressed due to the strong on-site Coulomb repulsion, so that the undoped cuprates are actually insulators with an energy gap $\sim$2 eV [16]. Furthermore, the spins in the copper sites interact via an exchange interaction mediated by the intermediate oxygen ions—a mechanism first described by Kramers and later applied by Anderson to transition metal oxides [17], known as "superexchange". Such an interaction is expected to promote magnetic ordering. It is easily seen that a spin alignment that facilitates nearest neighbor electron hopping is antiferromagnetic (see Fig. 2.1b), for in that case there is no Pauli principle restriction. This magnetic ordering has indeed been detected in all undoped cuprates.

As noted in the introduction, a number of complex phases emerge when charge carriers are doped into the CuO_2 planes. The following discussion of the phase diagram will be centered on the superconducting and pseudogap phases of *hole-doped* cuprates.

2.3 Cuprate Phase Diagram: Superconductivity and the Pseudogap

The passage from the Mott-insulating to the superconducting state is accomplished by introducing charge carriers into the system; specifically, into the CuO_2 planes. Indeed, an added positive charge (or hole) enters the spacer layers, increasing their overall charge by +1. To keep these layers neutral, the CuO_2 planes lose one electron, effectively leaving a hole behind which can now hop between planar sites [16]. At low temperatures, this hopping melts the AFM order and the system enters a phase best thought of as a disordered spin liquid [18]. Further doping at low temperatures brings the system to the superconducting state, which extends up to dopings of about 25%, with a maximum critical temperature around 15% doping. Above 25%, sufficiently far from the Mott insulator, the cuprates reach a phase more properly described as a Fermi liquid, with well-defined quasiparticles. Figure 2.2b shows a temperature vs. hole doping phase diagram deduced from ARPES measurements exclusively [19]. Except for the missing AFM phase near zero doping, this diagram is a comprehensive compendium of measurements done on different samples with dopings ranging from highly underdoped (UD) to highly overdoped (OD), and for temperatures between T_c and room temperature. d-wave superconductivity is realized in the gray dome-shaped region, while Fermi liquid behavior dominates the green region.

For doping levels above ~5%, the electron system develops a large Fermi surface. Figure 2.2a shows the form of this surface obtained from tight-binding [14] fits to ARPES data on Bi2212 in the normal state [20]. The bonding band has a minimum at Γ and maxima at Y, X. High resolution measurements on Pb-substituted Bi2212 have revealed that the Fermi surface in this compound consists of hole pockets centered at the corners $(\pm\pi/a, \pm\pi/a)$ which expand with hole-doping [21]. As we shall see in Chap. 5, the shape of the cuprate Fermi surface is important in this work because it supports scattering between quasiparticles in the antinodal regions of the Fermi surface separated by the commensurate wave-vector $\mathbf{Q} = (\pi/a, \pi/a)$ (blue arrow in Fig. 2.2a). Within the RPA approach adopted here, this enhanced scattering is ultimately responsible for the commensurate feature

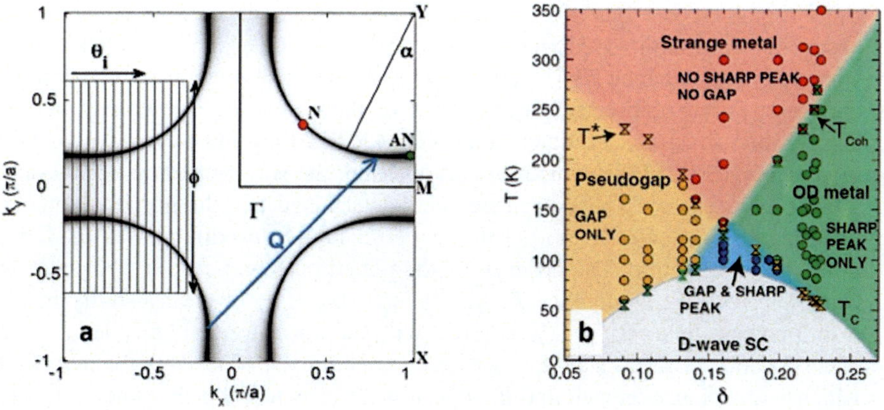

Fig. 2.2 (**a**) Fermi Surface obtained from tight-binding fits to normal state Bi2212 ARPES data. The node (N) and antinode (AN) of the superconducting gap are shown, as well as the slice (θ_i) and multiplexing (ϕ) angles (see Sects. 3.2 and 5.1). $\mathbf{Q} = (\pi/a, \pi/a)$ is the AFM ordering wave-vector. (**b**) ARPES-derived phase diagram. The circles are the points where spectra were collected. (Reprinted from Chatterjee, U. et al. Proceedings of the National Academy of Sciences 108, 9346–9349 (2011))

in the spin susceptibility and thereby the momentum anisotropy of the pairing interaction (see Sect. 2.4).

The superconducting state of the cuprates is not too different from that of conventional superconductors. Indeed, the charge carriers have an effective charge $2e$, reflecting the formation of Cooper pairs. Charge spectroscopies and conductivity measurements have also detected sharply peaked, long-lived excitations, consistent with Bogoliubov quasiparticles typical of BCS superconductors [22–26]. One important, unconventional attribute of the cuprates is the anisotropy (in momentum space) of the superconducting gap mentioned in the last section. Even though early ARPES [27–29], specific heat [30], penetration depth [31] and other measurements observed an anisotropic gap, they were all ultimately limited by their inability to detect its sign changes, so the issue was not settled until phase-sensitive experiments [32, 33] confirmed the d-wave ($\cos k_x - \cos k_y$) gap symmetry of the cuprates. This gap opens up at the Fermi Surface, has a maximum value at the antinode (AN) and vanishes at the node (N) along the BZ diagonals (Figs. 2.2a and 2.3).

Another key difference between cuprate and conventional superconductivity is found in the temperature dependence of the gap. ARPES experiments on Bi2212 have shown that the antinodal gap is approximately constant with temperature [34], in contrast with the mean-field behavior predicted by BCS. Even more striking is the observation that this gap persists above T_c with roughly the same magnitude, where the system is now in the *pseudogap* state, which we discuss next.

As the name suggests, this is a state where a gap opens in the single-particle excitation spectrum at a temperature T^* well above T_c. As shown in the yellow and blue regions in Fig. 2.2b, the pseudogap state persists through a large doping range, and extends up to room temperature near the AFM phase. This gap was initially inferred from NMR measurements, which showed a suppression of the low-energy spin response below T^* [35, 36]. Later optical conductivity measurements observed a drop in the c-axis (perpendicular to the CuO_2 planes) conductivity with temperature setting in at T^* [37]. The existence of an energy gap was inferred from these two probes indirectly, since neither of them measure the density of states itself. More direct observations came with ARPES [38] and Scanning Tunneling Microscopy (STM) [39], which directly measure the single-electron excitations as a function of energy. These measurements determined that the gapped excitations were incoherent (short-lived), so that a Fermi liquid description was inadequate in this doping regime. The absence of long-lived quasiparticles in this phase was consistent with the anomalous T^m-dependence ($m < 2$) of the resistivity inferred from transport data [40, 41], in contrast with the normal metal behavior $\rho \sim T^2$. ARPES data on Bi2212 also revealed the momentum dependence of the pseudogap, which at first glance seemed to follow the d-wave form found in the superconducting state. However, further experiments showed that the pseudogap actually vanished along arcs centered at the nodes of the gap function—the *Fermi arcs* [42]. The length of these arcs was later determined to be directly proportional to T/T^*, so that they evolved from four point nodes in the BZ at $T = 0$ to spanning the entire Fermi surface at $T = T^*$ [43]. Of particular importance to this work is the behavior of the pseudogap at the antinode in the BZ. At a given doping level, we now know

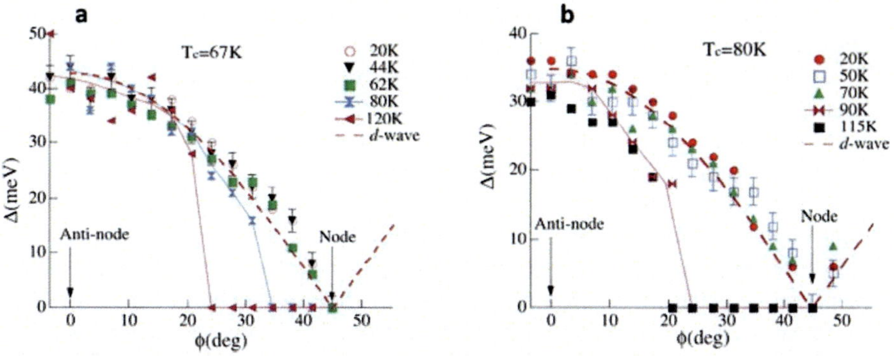

Fig. 2.3 (**a**) ARPES antinodal gap as a function of momentum and temperature for the underdoped Bi2212 thin film (UD67). (**b**) Same as (**a**), but for an optimally doped thin film (OP80). (Reprinted figure with permission from Kanigel, A. et al. Physical Review Letters 99, 157001 (2007). Copyright 2007 by the American Physical Society. https://doi.org/10.1103/PhysRevLett.99.157001)

it is roughly equal to the superconducting gap and constant in temperature up to T^*, where it "fills in" rather than closing [34]. As a function of doping, the antinodal pseudogap again follows the superconducting gap in that it becomes larger with underdoping [34]. These properties are summarized in Fig. 2.3.

2.4 Magnetism in the Cuprates and Superconductivity

In SF-induced superconductivity models, the momentum anisotropy of the purported pairing interaction

$$V_{\mathrm{SF}}(\mathbf{k}, \mathbf{k}', \Omega) = \frac{3}{2}U^2\chi(\mathbf{k} - \mathbf{k}', \Omega) = \frac{3}{2}U^2\chi(\mathbf{q}, \Omega) \tag{2.19}$$

is contained in the bosonic propagator χ (the spin susceptibility). The highly anisotropic momentum dependence of this propagator was first deduced from NMR measurements on the Yttrium-based cuprate $YBa_2Cu_3O_{6+x}$ (YBCO), which detected copper spin-lattice relaxation rates exceeding those of the oxygen ions by a factor of 20 [44]. This observation is rather counter-intuitive, since oxygen and copper are near neighbors with overlapping orbitals (see Fig. 2.1b), so one would expect them to interact strongly and thereby the Cu ions to "relax" the O ion. However, each O ion in the CuO_2 planes has a Cu ion on either side, so that the hyperfine coupling between the Cu spins and the O nuclear moment cancels out. More precisely, the hyperfine coupling between the O p-orbitals and the Cu d-orbitals is of the form $A(\mathbf{q}) = \cos\left(\frac{q\pi}{2Q}\right)$, which vanishes at $\mathbf{q} = \mathbf{Q} = (\pi, \pi)$. Since the O relaxation rate is proportional to the product of this hyperfine coupling and the magnetic density of states [45], then this latter quantity must have its spectral weight concentrated in a region of momentum space where $A(\mathbf{q})$ vanishes. Therefore, a suppressed oxygen relaxation rate implies a spin susceptibility peaked near $\mathbf{q} = \mathbf{Q}$.

Much more direct evidence for a spin susceptibility which is enhanced near the commensurate wave-vector $\mathbf{Q}$ has come from INS, which directly probes the imaginary part of the spin susceptibility, including its momentum and energy dependence. Indeed, INS data from the cuprates show a sharp enhancement of the spin susceptibility at the commensurate wave-vector below T_c [46–50]. This feature has been detected in several cuprate families [51] and, in the case of YBCO, it has been measured at various doping levels [52]. This "spin resonance" is a sharp peak located at $\mathbf{Q} = (\pi, \pi)$, whose energy (Ω_0) scales with T_c [52–54] and its intensity decreases with hole doping [52]. In underdoped YBCO, the resonance broadens and loses intensity as the sample is heated to the normal state [55, 56]. This peak—which we refer to as the commensurate feature, as it is no longer a sharp resonance—continues to decay and broaden until it finally disappears at the pseudogap temperature T^*. At optimal doping, INS has been so far unable to detect any magnetic signal in the normal state of the YBCO, despite its detection through other probes like NMR [44, 57]. However, the INS experiments of Xu et al. [58] have revealed magnetic excitations in optimally doped Bi2212 above T_c. Similarly to underdoped YBCO, the authors observed a precursor of the spin resonance with decreased intensity but located at roughly the same energy.

Aside from the commensurate response, INS has also revealed magnetic excitations above and below Ω_0 at incommensurate momenta. Indeed, yttrium, bismuth, and lanthanum-based cuprates show an excitation spectrum consisting of two branches emanating from the resonance coordinates ($\mathbf{Q}$, Ω_0) [59–63]. The so-called "upper branch" ($\Omega > \Omega_0$) is detected in all cuprates measured so far, whereas the "lower branch"($\Omega < \Omega_0$), which has been associated with stripe physics, appears to be material-specific. These two branches form what has come to be known as the famous "hourglass" dispersion of magnetic excitations in the superconducting cuprates.

Additional details regarding INS measurements in the cuprates will be deferred to Chap. 5, when we calculate the spin susceptibility from ARPES data. We close this chapter with a discussion of the implications of the momentum anisotropy of the function $\chi(\mathbf{q}, \Omega)$ for superconductivity.

In the BCS theory, the gap equation (Eq. 2.15) is solved for the simplest case of a constant, attractive interaction, which has the obvious appeal of rendering both sides of the equation positive. The pairing interaction in the spin fluctuation scenario given by Eq. 2.19 is repulsive and, in the case of the CuO_2 planes of the cuprates, has maxima near the two-dimensional BZ corners ($\mathbf{q} = \mathbf{Q} = (\pm\pi, \pm\pi)$). As only electrons near the Fermi surface are expected to participate in the pairing, $\epsilon_k \sim 0$ and we can approximately write $E_k \sim \Delta_k$. Moreover, since the momentum sum in Eq. 2.15 is over the first BZ and, as such, is invariant with respect to a (π, π) shift of the momentum variable, it can be restricted to a momentum region centered at $\mathbf{Q} = (\pi, \pi)$. As $V(\mathbf{k} - \mathbf{k}')$ is peaked at $\mathbf{k} - \mathbf{k}' = \mathbf{Q}$, the momentum sum in the gap equation is dominated by the term $\mathbf{k}' = \mathbf{k} - \mathbf{Q}$. Therefore, we have

$$\Delta_k \sim -\sum_{\omega_n} \frac{\Delta_{k-Q}}{\omega_n^2 + \Delta_{k-Q}^2} V_Q, \tag{2.20}$$

which admits a solution only if the gap changes sign when displaced through the diagonal vector **Q**. Thus, the SF pairing interaction directly implies d-wave pairing.

To gain physical insight as to how a purely repulsive interaction can give rise to Cooper pairing, we can begin by thinking of a negatively charged, static impurity in an electron gas. It has long been known that such an impurity induces spatial oscillations in the charge density—*Friedel oscillations* [6]—with wave-vector $2k_F$, where k_F is the Fermi momentum. If the impurity is placed at the origin, there are spatial coordinates where the electrostatic potential is negative, so that a second electron located at these coordinates will feel an attractive force to the impurity. The possible relevance of this phenomenon to a system where electrons pair not with a static impurity, but with each other, was first suggested by Kohn and Luttinger [64]. It is interesting to note that, like conventional superconductivity and magnetism, this mechanism relies on the Pauli principle. Indeed, a charged impurity in a *classical* electron gas will induce a charge density decaying exponentially with distance, but without a sign change, so pairing from repulsion would be impossible.

If we now imagine a two-dimensional, square lattice with spacing a (a model relevant to the cuprates, as discussed in Sect. 2.2) and seek the representation of V_{SF} in real space, we need only take a Fourier transform. In the bond directions, the Fourier transform of a function peaked at the BZ corner $(\pi/a, \pi/a)$ is a decaying oscillation with period $2a$. Therefore $V_{\mathrm{SF}}(\mathbf{r})$ is most attractive at $|\mathbf{r}| = a$, implying a pairing interaction between nearest neighbors in the square lattice [65]. Thus, the momentum dependence of the pairing interaction—peaked at high momenta—is consistent with the Friedel oscillation picture given above. Whereas in phonon-mediated superconductivity electrons avoid Coulomb repulsion by taking advantage of the non-locality of the pairing interaction in *time* (retardation), magnetically mediated pairing takes advantage of the non-locality of the interaction in *space* (near-neighbor coupling), although there may also be some retardation for this kind of interaction (see Chap. 5).

References

1. P.B. Allen, B. Mitrović, Theory of Superconducting T_c, in *Solid State Physics*, vol. 37 (Academic Press, 1983), pp. 1–92
2. H. Kamerlingh Onnes, Further experiments with liquid helium. C. On the change of electric resistance of pure metals at very low temperatures etc. IV. The resistance of pure mercury at helium temperatures. K. Ned. Akad. Wetensch. Proc. B Phys. Sci. **13**, 1274–1276 (1910)

3. W. Meissner, R. Ochsenfeld, Ein neuer Effekt bei Eintritt der Supraleitfähigkeit. Naturwissenschaften **21**(44), 787–788 (1933)
4. J. Bardeen, D. Pines, Electron-phonon interaction in metals. Phys. Rev. **99**(4), 1140–1150 (1955)
5. J. Bardeen, L.N. Cooper, J.R. Schrieffer, Theory of superconductivity. Phys. Rev. **108**(5), 1175–1204 (1957)
6. A.L. Fetter, J.D. Walecka, *Quantum Theory of Many-Particle Systems* (McGraw-Hill, Boston, 1971)
7. E.B. Hansen, Infinite conductivity of ordinary and gapless superconductors. Physica **39**(2), 271–292 (1968)
8. F. London, H. London, F.A. Lindemann, The electromagnetic equations of the supraconductor. Proc. R. Soc. Lond. A Math. Phys. Sci. **149**(866), 71–88 (1935)
9. N. Plakida, *High-Temperature Cuprate Superconductors: Experiment, Theory, and Applications*. Springer Series in Solid-State Sciences (Springer, Berlin, 2010)
10. J.A. Slezak, J. Lee, M. Wang, K. McElroy, K. Fujita, B.M. Andersen, P.J. Hirschfeld, H. Eisaki, S. Uchida, J.C. Davis, Imaging the impact on cuprate superconductivity of varying the interatomic distances within individual crystal unit cells. Proc. Natl. Acad. Sci. U. S. Am. **105**(9), 3203–3208 (2008)
11. D. Fausti, R.I. Tobey, N. Dean, S. Kaiser, A. Dienst, M.C. Hoffmann, S. Pyon, T. Takayama, H. Takagi, A. Cavalleri. Light-induced superconductivity in a stripe-ordered cuprate. Science **331**(6014), 189–191 (2011)
12. R. Mankowsky, A. Subedi, M. Först, S.O. Mariager, M. Chollet, H.T. Lemke, J.S. Robinson, J.M. Glownia, M.P. Minitti, A. Frano, M. Fechner, N.A. Spaldin, T. Loew, B. Keimer, A. Georges, A. Cavalleri, Nonlinear lattice dynamics as a basis for enhanced superconductivity in $YBa_2Cu_3O_{6.5}$. Nature **516**, 71 (2014)
13. P.W. Anderson, *The Theory of Superconductivity in the High-Tc Cuprate Superconductors* (Princeton University Press, 1997)
14. N.W. Ashcroft, N.D. Mermin, *Solid State Physics* (Brooks/Cole, Australia, 1976)
15. N.F. Mott, The basis of the electron theory of metals, with special reference to the transition metals. Proc. Phys. Soc. Sect. A **62**(7), 416–422 (1949)
16. M.R. Norman, D. Pines, C. Kallin, The pseudogap: friend or foe of high T_c? Adv. Phys. **54**(8), 715–733 (2005)
17. P.W. Anderson, Antiferromagnetism. Theory of superexchange interaction. Phys. Rev. **79**(2), 350–356 (1950)
18. M.R. Norman, High-temperature superconductivity–magnetic mechanisms, in *Handbook of Magnetism and Advanced Magnetic Materials* (Wiley, 2007)
19. U. Chatterjee, D. Ai, J. Zhao, S. Rosenkranz, Adam Kaminski, H. Raffy, Z. Li, K. Kadowaki, M. Randeria, M.R. Norman, J.C. Campuzano, Electronic phase diagram of high-temperature copper oxide superconductors. Proc. Natl. Acad. Sci. **108**(23), 9346–9349 (2011)
20. M.R. Norman, Linear response theory and the universal nature of the magnetic excitation spectrum of the cuprates. Phys. Rev. B **75**(18), 184514 (2007)
21. A.A. Kordyuk, S.V. Borisenko, M.S. Golden, S. Legner, K.A. Nenkov, M. Knupfer, J. Fink, H. Berger, L. Forró, R. Follath, Doping dependence of the Fermi surface in $(Bi,Pb)_2Sr_2CaCu_2O_{8+\delta}$. Phys. Rev. B **66**(1), 014502 (2002)
22. D.A. Bonn, P. Dosanjh, R. Liang, W.N. Hardy, Evidence for rapid suppression of quasiparticle scattering below T_c in $YBa_2Cu_3O_{7-\delta}$. Phys. Rev. Lett. **68**(15), 2390–2393 (1992)
23. K. Krishana, J.M. Harris, N.P. Ong, Quasiparticle mean free path in $YBa_2Cu_3O_7$ measured by the thermal hall conductivity. Phys. Rev. Lett. **75**(19), 3529–3532 (1995)
24. A. Kaminski, J. Mesot, H. Fretwell, J.C. Campuzano, M.R. Norman, M. Randeria, H. Ding, T. Sato, T. Takahashi, T. Mochiku, K. Kadowaki, H. Hoechst, Quasiparticles in the superconducting state of $Bi_2Sr_2CaCu_2O_{8+\delta}$. Phys. Rev. Lett. **84**(8), 1788–1791 (2000)
25. J.C. Campuzano, H. Ding, M.R. Norman, M. Randeria, A.F. Bellman, T. Yokoya, T. Takahashi, H. Katayama-Yoshida, T. Mochiku, K. Kadowaki, Direct observation of particle-hole mixing in the superconducting state by angle-resolved photoemission. Phys. Rev. B **53**(22), R14737–R14740 (1996)

26. H. Matsui, T. Sato, T. Takahashi, S.C. Wang, H.B. Yang, H. Ding, T. Fujii, T. Watanabe, A. Matsuda, BCS-like Bogoliubov quasiparticles in high-T_c superconductors observed by angle-resolved photoemission spectroscopy. Phys. Rev. Lett. **90**(21), 217002 (2003)
27. H. Ding, J.C. Campuzano, K. Gofron, C. Gu, R. Liu, B.W. Veal, G. Jennings, Gap anisotropy in $Bi_2Sr_2CaCu_2O_{8+\delta}$ by ultrahigh-resolution angle-resolved photoemission. Phys. Rev. B **50**(2), 1333–1336 (1994)
28. H. Ding, J.C. Campuzano, A.F. Bellman, T. Yokoya, M.R. Norman, M. Randeria, T. Takahashi, H. Katayama-Yoshida, T. Mochiku, K. Kadowaki, G. Jennings, Momentum dependence of the superconducting gap in $Bi_2Sr_2CaCu_2O_{8+\delta}$. Phys. Rev. Lett. **75**(7), 1425–1425 (1995)
29. Z.X. Shen, D.S. Dessau, B.O. Wells, D.M. King, W.E. Spicer, A.J. Arko, D. Marshall, L.W. Lombardo, A. Kapitulnik, P. Dickinson, S. Doniach, J. DiCarlo, T. Loeser, C.H. Park, Anomalously large gap anisotropy in the a-b plane of $Bi_2Sr_2CaCu_2O_{8+\delta}$. Phys. Rev. Lett. **70**(10), 1553–1556 (1993)
30. D.M. Ginsberg, *Physical Properties of High Temperature Superconductors III* (World Scientific, 1992)
31. J. Annett, N. Goldenfeld, S.R. Renn, Interpretation of the temperature dependence of the electromagnetic penetration depth in $YBa_2Cu_3O_{7-\delta}$. Phys. Rev. B **43**(4), 2778–2782 (1991)
32. D.A. Wollman, D.J. Van Harlingen, W.C. Lee, D.M. Ginsberg, A.J. Leggett, Experimental determination of the superconducting pairing state in YBCO from the phase coherence of YBCO-Pb dc SQUIDs. Phys. Rev. Lett. **71**(13), 2134–2137 (1993)
33. C.C. Tsuei, J.R. Kirtley, C.C. Chi, L.S. Yu-Jahnes, A. Gupta, T. Shaw, J.Z. Sun, M.B. Ketchen, Pairing symmetry and flux quantization in a tricrystal superconducting ring of $YBa_2Cu_3O_{7-\delta}$. Phys. Rev. Lett. **73**(4):593–596 (1994)
34. A. Kanigel, U. Chatterjee, M. Randeria, M.R. Norman, S. Souma, M. Shi, Z.Z. Li, H. Raffy, J.C. Campuzano, Protected nodes and the collapse of fermi arcs in high-T_c cuprate superconductors. Phys. Rev. Lett. **99**(15), 157001 (2007)
35. W.W. Warren, R.E. Walstedt, G.F. Brennert, R.J. Cava, R. Tycko, R.F. Bell, G. Dabbagh, Cu spin dynamics and superconducting precursor effects in planes above T_c in $YBa_2Cu_3O_{6.7}$. Phys. Rev. Lett. **62**(10), 1193–1196 (1989)
36. H. Alloul, T. Ohno, P. Mendels, ^{89}NMR evidence for a fermi-liquid behavior in $YBa_2Cu_3O_{6+x}$. Phys. Rev. Lett. **63**(16), 1700–1703 (1989)
37. C.C. Homes, T. Timusk, R. Liang, D.A. Bonn, W.N. Hardy, Optical conductivity of c-axis oriented $YBa_2Cu_3O_{6.7}$: Evidence for a pseudogap. Phys. Rev. Lett. **71**(10), 1645–1648 (1993)
38. H. Ding, T. Yokoya, J.C. Campuzano, T. Takahashi, M. Randeria, M.R. Norman, T. Mochiku, K. Kadowaki, J. Giapintzakis, Spectroscopic evidence for a pseudogap in the normal state of underdoped high-T_c superconductors. Nature **382**(6586), 51–54 (1996)
39. Ch. Renner, B. Revaz, J.Y. Genoud, K. Kadowaki, Ø. Fischer, Pseudogap precursor of the superconducting gap in under- and overdoped $Bi_2Sr_2CaCu_2O_{8+\delta}$. Phys. Rev. Lett. **80**(1), 149–152 (1998)
40. S.H. Naqib, J.R. Cooper, J.L. Tallon, C. Panagopoulos, Temperature dependence of electrical resistivity of high-T_c cuprates—from pseudogap to overdoped regions. Physica C Superconductivity **387**(3), 365–372 (2003)
41. A. Yamamoto, N. Takeshita, C. Terakura, Y. Tokura, High pressure effects revisited for the cuprate superconductor family with highest critical temperature. Nat. Commun. **6**(1), 8990 (2015)
42. M.R. Norman, H. Ding, M. Randeria, J.C. Campuzano, T. Yokoya, T. Takeuchi, T. Takahashi, T. Mochiku, K. Kadowaki, P. Guptasarma, D.G. Hinks, Destruction of the Fermi surface in underdoped high-T_c superconductors. Nature **392**, 157 (1998)
43. A. Kanigel, M.R. Norman, M. Randeria, U. Chatterjee, S. Souma, A. Kaminski, H.M. Fretwell, S. Rosenkranz, M. Shi, T. Sato, T. Takahashi, Z.Z. Li, H. Raffy, K. Kadowaki, D. Hinks, L. Ozyuzer, J.C. Campuzano, Evolution of the pseudogap from Fermi arcs to the nodal liquid. Nat. Phys. **2**(7), 447–451 (2006)
44. P.C. Hammel, M. Takigawa, R.H. Heffner, Z. Fisk, K.C. Ott, Spin dynamics at oxygen sites in $YBa_2Cu_3O_7$. Phys. Rev. Lett. **63**(18), 1992–1995 (1989)

45. M. Takigawa, A.P. Reyes, P.C. Hammel, J.D. Thompson, R.H. Heffner, Z. Fisk, K.C. Ott, Cu and O NMR studies of the magnetic properties of $YBa_2Cu_3O_{6.63}$ (T_c=62 K). Phys. Rev. B **43**(1), 247–257 (1991)
46. J. Rossat-Mignod, L.P. Regnault, C. Vettier, P. Bourges, P. Burlet, J. Bossy, J.Y. Henry, G. Lapertot, Neutron scattering study of the $YBa_2Cu_3O_{6+x}$ system. Physica C Superconductivity **185–189**, 86–92 (1991)
47. H.F. Fong, B. Keimer, P.W. Anderson, D. Reznik, F. Doğan, I.A. Aksay, Phonon and magnetic neutron scattering at 41 meV in $YBa_2Cu_3O_7$. Phys. Rev. Lett. **75**(2), 316–319 (1995)
48. H.F. Fong, P. Bourges, Y. Sidis, L.P. Regnault, A. Ivanov, G.D. Gu, N. Koshizuka, B. Keimer, Neutron scattering from magnetic excitations in $Bi_2Sr_2CaCu_2O_{8+\delta}$. Nature **398**(6728), 588–591 (1999)
49. H.A. Mook, M. Yethiraj, G. Aeppli, T.E. Mason, T. Armstrong, Polarized neutron determination of the magnetic excitations in $YBa_2Cu_3O_7$. Phys. Rev. Lett. **70**(22), 3490–3493 (1993)
50. M.K. Chan, C.J. Dorow, L. Mangin-Thro, Y. Tang, Y. Ge, M.J. Veit, G. Yu, X. Zhao, A.D. Christianson, J.T. Park, Y. Sidis, P. Steffens, D.L. Abernathy, P. Bourges, M. Greven, Commensurate antiferromagnetic excitations as a signature of the pseudogap in the tetragonal high-T_c cuprate $HgBa_2CuO_{4+\delta}$. Nat. Commun. **7**(1), 1–7 (2016)
51. G. Yu, Y. Li, E.M. Motoyama, M. Greven, A universal relationship between magnetic resonance and superconducting gap in unconventional superconductors. Nat. Phys. **5**(12), 873–875 (2009)
52. B. Keimer, H.F. Fong, S.H. Lee, D.L. Millius, I.A. Aksay, Doping dependence of the magnetic resonance peak in $YBa_2Cu_3O_{6+x}$. Physica C Superconductivity **282–287**, 232–235 (1997)
53. H. He, Y. Sidis, P. Bourges, G.D. Gu, A. Ivanov, N. Koshizuka, B. Liang, C.T. Lin, L.P. Regnault, E. Schoenherr, B. Keimer, Resonant spin excitation in an overdoped high temperature superconductor. Phys. Rev. Lett. **86**(8), 1610–1613 (2001)
54. L. Capogna, B. Fauqué, Y. Sidis, C. Ulrich, P. Bourges, S. Pailhès, A. Ivanov, J.L. Tallon, B. Liang, C.T. Lin, A.I. Rykov, B. Keimer, Odd and even magnetic resonant modes in highly overdoped $Bi_2Sr_2CaCu_2O_{8+\delta}$. Phys. Rev. B **75**(6), 060502 (2007)
55. P. Dai, H.A. Mook, S.M. Hayden, G. Aeppli, T.G. Perring, R.D. Hunt, F. Doğan, The magnetic excitation spectrum and thermodynamics of high-T_c superconductors. Science **284**(5418), 1344–1347 (1999)
56. H.F. Fong, B. Keimer, D.L. Milius, I.A. Aksay, Superconductivity-induced anomalies in the spin excitation spectra of underdoped $YBa_2Cu_3O_{6+x}$. Phys. Rev. Lett. **78**(4), 713–716 (1997)
57. A. Uldry, M. Mali, J. Roos, P.F. Meier, Anisotropy of the antiferromagnetic spin correlations in the superconducting state of $YBa_2Cu_3O_7$ and $YBa_2Cu_3O_8$. J. Phys. Condens. Matter **17**(46), L499–L505 (2005)
58. G. Xu, G.D. Gu, M. Hücker, B. Fauqué, T.G. Perring, L.P. Regnault, J.M. Tranquada, Testing the itinerancy of spin dynamics in superconducting $Bi_2Sr_2CaCu_2O_{8+\delta}$. Nat. Phys. **5**(9), 642–646 (2009)
59. S.W. Cheong, G. Aeppli, T.E. Mason, H. Mook, S.M. Hayden, P.C. Canfield, Z. Fisk, K.N. Clausen, J.L. Martinez, Incommensurate magnetic fluctuations in $La_{2-x}Sr_xCuO_4$. Phys. Rev. Lett. **67**(13), 1791–1794 (1991)
60. H.A. Mook, P. Dai, S.M. Hayden, G. Aeppli, T.G. Perring, F. Doğan, Spin fluctuations in $YBa_2Cu_3O_{6.6}$. Nature **395**(6702), 580–582 (1998)
61. V. Hinkov, P. Bourges, S. Pailhès, Y. Sidis, A. Ivanov, C.D. Frost, T.G. Perring, C.T. Lin, D.P. Chen, B. Keimer, Spin dynamics in the pseudogap state of a high-temperature superconductor. Nat. Phys. **3**(11), 780–785 (2007)
62. B. Fauqué, Y. Sidis, L. Capogna, A. Ivanov, K. Hradil, C. Ulrich, A.I. Rykov, B. Keimer, P. Bourges, Dispersion of the odd magnetic resonant mode in near-optimally doped $Bi_2Sr_2CaCu_2O_{8+\delta}$. Phys. Rev. B **76**(21), 214512 (2007)
63. O.J. Lipscombe, B. Vignolle, T.G. Perring, C.D. Frost, S.M. Hayden, Emergence of coherent magnetic excitations in the high temperature underdoped $La_{2-x}Sr_xCuO_4$ superconductor at low temperatures. Phys. Rev. Lett. **102**(16), 167002 (2009)

64. W. Kohn, J.M. Luttinger, New mechanism for superconductivity. Phys. Rev. Lett. **15**(12), 524–526 (1965)
65. D.J. Scalapino, A common thread: The pairing interaction for unconventional superconductors. Rev. Mod. Phys. **84**(4), 1383–1417 (2012)

Chapter 3
Angle-Resolved Photoemission Spectroscopy

Abstract Owing to the approximate translational invariance of the Hamiltonian experienced by electrons in crystals, electronic excitations in solids are most conveniently studied in momentum space. It is therefore desirable to probe observables like single-particle energies as functions of good quantum numbers like (crystal) momentum. This is precisely the kind of measurement afforded by Angle-Resolved Photoemission Spectroscopy (ARPES), one of the most incisive microscopic probes employed in condensed matter physics. This chapter covers the basic principles of ARPES, as well as its connection with a quantity of fundamental theoretical interest in many-body physics, the Green's function.

As mentioned in the introduction, in this work we employ photoemission data to construct all the quantities entering the generalized gap equation. This chapter covers the subject of Angle-Resolved Photoemission Spectroscopy (ARPES). Sections 3.1 and 3.2 cover the general principles and some of the capabilities and limitations of the technique. Section 3.3 is devoted to the conservation laws that make this technique useful in the study of quantum systems. In Sect. 3.4, the important connection between photoemission spectra and the "spectral function" is addressed. Finally, Sect. 3.5 summarizes the different visualizations of ARPES data that are commonly used.

3.1 Basic Description

The name "Angle-Resolved Photoemission Spectroscopy" (ARPES) contains three components: (1) *Photoemission* describes the basic physical phenomenon governing the technique; i.e. the photoelectric effect. When a sufficiently energetic photon interacts with a given material, its energy is absorbed and imparted to an electron which may escape the sample (out to the "vacuum") with some residual kinetic energy. For the light sources typically used in practice, there are very many such photoemission events, which collectively give rise to a "photocurrent" (current of

F. Restrepo, *Angle-Resolved Photoemission Spectroscopy Study of Spin Fluctuations in the Cuprate Superconductors*, Springer Theses,
https://doi.org/10.1007/978-3-031-10979-9_3

photo-emitted electrons). (2) If it were somehow possible to collect these electrons and classify them according to their kinetic *energy*, one would obtain a distribution, i.e. a plot of detected electrons as a function of kinetic energy. The study of such distributions is called *spectroscopy*. (3) If, in addition, the emitted electrons (or photoelectrons) are sorted out according to the *direction* in which they were emitted, one then has angular resolution. Therefore, the photoelectric effect where the kinetic energy and emission angle of the photoelectrons are known is called ARPES.

Angle and energy resolution are usually achieved in practice with the aid of a Spherical Deflection Analyzer (SDA). Other types of analyzers have been developed throughout the years, especially during the early days of angle-resolved photoemission [1]. However, because the SDA is currently the most widely used spectrometer and because it is the one used in this work, only this kind will be described below.

A typical SDA consists of two concentric half-spherical surfaces of radii R_1 and R_2, where $R_2 > R_1$. Photoelectrons enter the analyzer through a narrow (in our case, vertical) *entrance slit*. The two shells are held at voltages V_1 and V_2 ($V_1 > V_2$) such that a potential difference ΔV is established between them (corresponding to an electric field in the positive radial direction), which forces the photoelectrons to follow near-circular trajectories. Electrons follow these trajectories until they reach the detector system. The plane containing the sample, the entrance slit, and the detector is the *energy dispersive* plane (Fig. 3.1a and b).

Electrons entering the slit with high kinetic energy will describe larger circumferences in this plane than those entering with lower kinetic energy. The detector assembly essentially acts as a two-dimensional grid which registers the electrons' final positions. This two-dimensional array is oriented in such a way that one of its axes (say, x) is contained in the energy dispersive plane, so that electrons registered at different x positions have different kinetic energies (Fig. 3.1b). The other dimension of the detector (say, y) is used for "angular dispersion" (or *multiplexing*). Having a vertical entrance slit means only a vertical "slice" of the

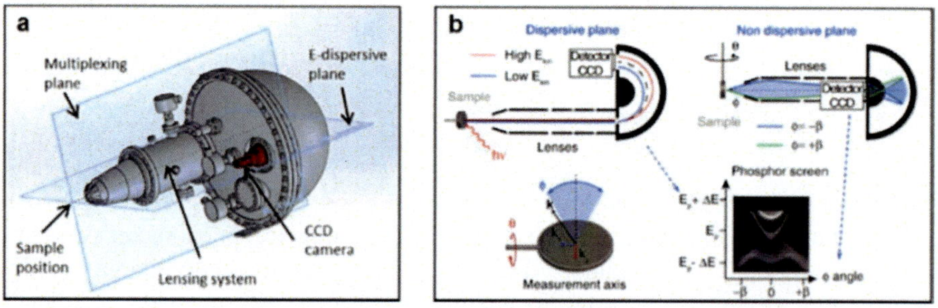

Fig. 3.1 (**a**) SCIENTA R4000 hemispherical analyzer. The energy-dispersive and multiplexing planes are shown, along with the approximate positions of the entrance slit, sample, and pre-entrance slit lensing system. (**b**) Schematic projections of a SDA on two planes: multiplexing plane and energy-dispersive plane. (Reprinted from Bianchi, M. et al. Semiconductor Science and Technology 27, 124001 (2012) ©IOP Publishing. Reproduced with permission. All rights reserved. https://doi.org/10.1088/0268-1242/27/12/124001)

photoemission solid angle is collected by the analyzer. Because of the finite length of the entrance slit, only electrons within an "acceptance angle" ϕ will enter the analyzer system. Upon their entrance, these electrons are straying away from the optical axis, so an electrostatic lensing system is in place to focus them onto the entrance slit. Therefore, when the photoelectrons enter the SDA proper, they do so with different incidence angles and at different points along the slit. These different entrance conditions are directly mapped onto the detector; i.e. electrons entering through different points in the slit will be registered at different points in the y detector dimension. To summarize, the x and y axes of the detector classify electrons according to their energy and emission angle, respectively. The scales of these axes can be calibrated in such a way that one can effectively map the real-space, two-dimensional intensity map registered by the detector onto an energy versus photoemission angle map. As shown in Sect. 3.3, the latter quantity can be directly related to the photoelectron momentum, so that the above mapping actually yields the dispersion relation $\varepsilon(\mathbf{k})$ as an intensity map, a very useful piece of information.

3.2 Limitations

There are a number of limitations to this technique, and it is important to understand them before analyzing ARPES data.

The two-dimensional nature of the detector implies that, if one insists on reserving one dimension for energy, there is only one dimension left for momentum; i.e. the detector only does multiplexing along the axis parallel to the entrance slit, so that one actually measures $\varepsilon(k_y)$. However, angular dispersion along the perpendicular direction can be realized by rotating the sample itself through angle θ in Fig. 3.1b. More precisely, for every value of θ (say, θ_i), one can obtain a dispersion relation of the form $\varepsilon^i(\theta_i, k_y) \sim \varepsilon^i(k_x^i, k_y)$. Rotating the sample through several angles θ_i effectively yields $\varepsilon(k_x, k_y)$. So, in practice, this drawback can be overcome. (More details in Sect. 4.3.)

Now, the question rises as to how to measure k_z, without which the full electron momentum vector is unknown. The sample has to possess a surface from which electrons escape. A macroscopic sample may be assumed to extend infinitely along the plane of the surface. Since the samples considered are crystalline, the Hamiltonian describing the system is invariant through lattice vector translations with an associated conservation of "crystal momentum" [2].[1] This invariance only holds in the plane parallel to the surface; in the normal direction this symmetry is severely broken by the presence of the surface. Therefore, in going from inside the sample to the vacuum, the momentum in the perpendicular direction (k_z) may not be conserved, whereas the in-plane components are. This non-conservation, imposed by the laws of quantum mechanics, expresses a fundamental drawback,

[1] From now on, the term "momentum" will be used instead of "crystal momentum" unless we refer to electrons outside the sample, where they can be described by their "real" momentum.

and it has very important consequences, for, as will be discussed in the next section, conservation laws imply a direct correspondence between what the detector registers and the electronic behavior inside the sample. Most practitioners (the author included) avoid this problem altogether by studying samples whose electronic structure is such that the single-particle energies are approximately independent of one of the momentum components. Such materials are dubbed *two-dimensional*.

An obvious limitation on any spectroscopy is resolution; in the present case, energy and momentum resolutions. The total energy and momentum resolution functions of the spectra are given by the convolution of the detector and light source (synchrotron, LASER, or Heumlium lamp) resolution functions. For instance, typical total resolutions for ultraviolet (UV) synchrotron radiation are about 15 meV for energy, 0.01 $Å^{-1}$ for momentum in the multiplexing direction and 0.02 $Å^{-1}$ in the direction perpendicular. Details regarding the instrumental resolution in this work can be found in Sect. 4.6.

For completeness, we should mention what is arguably the most challenging limitation of this technique: surface sensitivity. Photoexcited electrons have escape depths of the order of 5 Å (for an estimate of this quantity for Bi2212, see [3]), which means that only the electronic behavior near the sample surface is probed. Consequently, the sample surface must be kept as clean as possible; otherwise, adsorbates may accumulate on it and scatter the outgoing electrons, which will then loose the energy-momentum information they were bringing from within the sample. So far, the best known way to accomplish this is by keeping the sample in Ultra-High Vacuum (UHV); that is, at pressures in the 10^{-11} Torr range (see Sect. 4.2). Surface sensitivity may pose the more serious problem of *surface states* [1]. These are states that form in the vicinity of the surface, where electrons experience a different electrostatic environment from those in the bulk. Fortunately, as mentioned in Sect. 2.2, the compound studied in this work, Bi2212, cleaves very easily and the surface environment is highly representative of the bulk. Hence, the spectra are not plagued by surface states.

3.3 Conservation Laws

The energies and momenta recorded by the detector are meaningful only insofar as they can be directly related to the corresponding quantities inside the sample. This correspondence in turn is only possible due to conservation of energy and momentum. Energy conservation is concisely expressed through Einstein's well-known equation describing the photoelectric effect [4]:

$$h\nu = E_{kin} + \phi + E_B \tag{3.1}$$

The energy of the incident photon is invested in promoting the electron from its initial quantum state up to the Fermi level (*binding energy* E_B), bringing it to the surface (*work function* ϕ), and giving it some kinetic energy to make it to the detector (E_{kin}). Here, one is particularly interested in the binding energy and its momentum dependence, for this quantity encodes valuable information about the quantum mechanical system and its interactions, whether they be between electrons themselves or between electrons and different collective modes.

Accurate determination of the Fermi energy is essential in any electron spectroscopy, since energies are most conveniently measured with respect to this reference. In the case of superconductors with a gapped density of states near the Fermi surface, knowledge of E_F is imperative, since the spectroscopic gap ($\sim$ 40 meV for Bi2212) is measured with respect to this value. Thermodynamics suggests a solution; one simply places the sample in electrical contact with a metal with a clearly discernible Fermi edge (like polycrystalline Au) and uses this edge as a reference. This works because two systems (sample and Au) which are free to exchange particles (electrons) do so until their chemical potentials (or Fermi energies, neglecting small corrections at finite temperatures) are equal. For an initial state with energy E_F (so that $E_B = 0$), Eq. 3.1 gives the work function $\phi(\mu) = \phi = h\nu - E_{kin}(\mu)$, where both quantities on the right hand side are known. Hence, $h\nu$ is set by the experiment, the detector measures E_{kin}, and ϕ can determined; thus, Einstein's equation can be solved for E_B.

As indicated above, only the in-plane component of the photoelectron momentum (labeled $k_{||}$) is conserved as it crosses the surface. Outside the sample (i.e. in vacuum), there is full translational symmetry and the electronic states are labeled by their "real" momentum. If an electron is emitted with total momentum (k) and emission polar angle ϕ (in a coordinate system where the z axis is normal to the sample surface), then the in-plane momentum is $k \sin\phi$. But, in vacuum, k and E_{kin} obey the simple "free-electron" dispersion relation $E_{kin} = \frac{\hbar^2 k^2}{2m}$, where m is the electron mass. Hence,

$$k_{||} = \frac{1}{\hbar}\sqrt{2mE_{kin}} \sin\phi \tag{3.2}$$

There exist several methods to reconstruct the dispersion along the k_z direction, some of which are discussed in [1]. Alternatively, one may concentrate on samples which are known to exhibit little or no k_z dispersion, as mentioned in Sect. 3.2.

Note that conservation of (parallel) momentum has only been invoked in the vicinity of the surface; according to the above, the measured photoelectron momentum is only guaranteed to yield the momentum of the electron just before it crossed the surface to the vacuum. However, we are interested in the momentum of the electron in the initial state, before it propagated to the surface, and there is no guarantee that these momenta are equal. On one hand, we have completely ignored the incident photon momentum. Fortunately, this contribution is only a fraction of a percent of the BZ size,[2] so it can be safely ignored in all but the highest

[2] The ratio of the photon to electron momentum is $\frac{p_\gamma}{p_e} = \frac{E/c}{h/a} = \frac{Ea}{hc}$. For a 10 eV photon (a reasonable value for UV photoemission) incident on a crystal with lattice parameter $a = 5$ Å (also a reasonable value), this ratio is 0.004.

photon energy experiments (not considered in this work). On the other hand, there is a more important correction to the electron momentum coming from its possible interactions on its way to the surface. For example, in the cuprate Bi2212, exiting electrons are sensitive to an incommensurate modulation of the lattice [5], giving rise to well-known *superlattice* bands (see data presented in Chap. 5). In this case, however, the effect is well known and one can subtract these artificial features from the main signal without difficulty.

3.4 The Spectral Function

Figure 3.2 shows an example of a "slice" corresponding to an angle θ_i (the data were actually collected from the High-T_c cuprate Bi2212, but the details are not important at the moment). One axis is labeled ω, corresponding to the binding energy measured with respect to the Fermi energy. The other is labeled k (corresponding to the multiplexed angles ϕ). The third dimension is just intensity (or number of electrons registered). The lower part of the figure shows two perpendicular cuts through the dispersion; one for $\omega = 0$ and one for $k = k_F$. Cuts for fixed energy and momentum are called Momentum Distribution Curves (MDCs) and

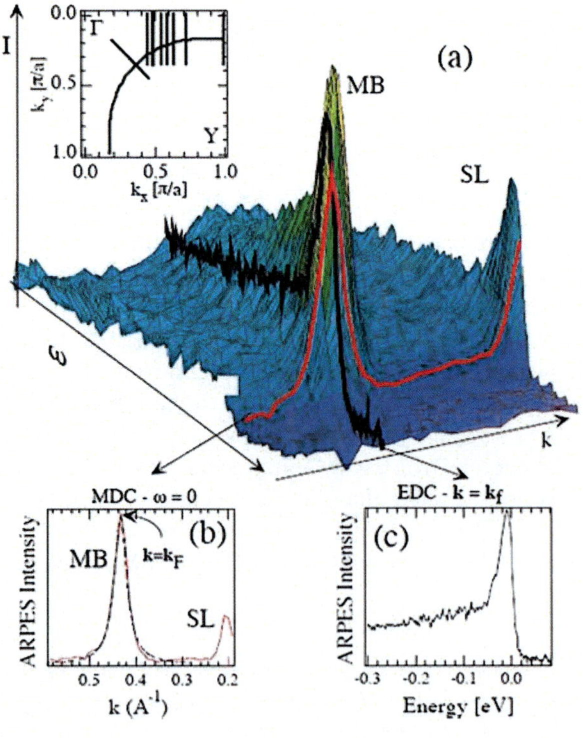

Fig. 3.2 (**a**) Intensity versus momentum and binding energy for a given θ_i. Such an intensity map corresponds to the (renormalized) dispersion ε_k, where the momentum variable is allowed to run over one dimension. (**b**) MDC at the Fermi energy ($\omega = 0$). (**c**) EDC taken at the Fermi momentum ($k = k_F$) (Reprinted figure with permission from Kaminski, A. et al. Physical Review Letters 86, 1070 - 1073 (2001). Copyright 2001 by the American Physical Society. https://doi.org/10.1103/PhysRevLett.86.1070)

Energy Distribution Curves (EDCs), respectively. These are just as important as the dispersion $\varepsilon(\mathbf{k})$, for there are good reasons to believe they are different projections of a quantity of fundamental theoretical interest: the *spectral function* $A(\mathbf{k}, \omega)$.

The details behind the spectral function interpretation of ARPES intensity are beyond the scope of this work, so we limit ourselves to mentioning a few highlights in the history of its development. Notable approaches to the theory of photoemission include that of Mahan [6], based on scattering theory, Ashcroft and Schaich's [7], based on a quadratic response formalism (from which they obtained results identical to Mahan's), and later Pendry's calculations including matrix element and hole lifetime effects [8]. Perhaps Hedin and Lundqvist drew a more direct connection between photoemission and the spectral function [9], based on Fermi's Golden Rule. Their work originally focused on metals excited with X-Rays, where photoelectrons are emitted with high enough energy that their interaction with the hole left behind is negligible, an assumption commonly referred to as the *sudden approximation*.

First-principles approaches like the ones mentioned above were complemented more recently by the phenomenological study of Randeria et al. [10], whose work is of special interest here because they used ARPES data from the cuprate Bi2212. Assuming the measured spectrum could be modeled by

$$I(\mathbf{k}, \omega) = I_0 f(\omega) A(\mathbf{k}, \omega) \tag{3.3}$$

and that the prefactor $I_0(\mathbf{k})$ -a matrix element for the photon-induced transition- was independent of T and ω, they were able to verify that the integral $\int d\omega I(\mathbf{k_F}, \omega)$ was independent of temperature, a sum rule which follows from the general properties of $A(\mathbf{k}, \omega)$ and which *must* be satisfied if Eq. 3.3 is to hold. ($f(\omega)$ is the Fermi function with $\mu = 0$.)

We close this section with a brief explanation of the physical meaning of the spectral function, defined as [10]

$$A(\mathbf{k}, \omega) = A^+(\mathbf{k}, \omega) + A^-(\mathbf{k}, \omega), \tag{3.4}$$

with

$$A^+(\mathbf{k}, \omega) = \sum_{n,m} |\langle n|c^\dagger_{\mathbf{k}}|m\rangle|^2 \frac{e^{-\beta E_m}}{Z} \delta(\omega + E_m - E_n) \tag{3.5}$$

$$A^-(\mathbf{k}, \omega) = \sum_{n,m} |\langle n|c_{\mathbf{k}}|m\rangle|^2 \frac{e^{-\beta E_m}}{Z} \delta(\omega + E_n - E_m). \tag{3.6}$$

In these expressions, Z is the partition function, $\beta = 1/k_B T$, and the states $|n\rangle$ and $|m\rangle$ are eigenstates of the Hamiltonian with energies E_n and E_m. To appreciate the physical significance of this spectral function, let us first consider the $A^-(\mathbf{k}, \omega)$ term. This term consists of an infinite sum of delta peaks, each weighted by a statistical factor and a scalar $|\langle n|c_{\mathbf{k}}|m\rangle|^2$. Now, this scalar is the inner product of the states $c_{\mathbf{k}}|m\rangle$ and $|n\rangle$. Assuming the state $|m\rangle$ represents an N-particle system, the state $c_{\mathbf{k}}|m\rangle$ belongs to the subspace of Fock space with $N-1$ particles, so this ket represents the state of the remaining $(N-1)$-particle system just after a particle with momentum $\mathbf{k}$ has been removed from state $|m\rangle$. However, this state is not an eigenstate of the Hamiltonian *in general*. Instead, it can be expressed as a linear combination of such eigenstates in the $(N-1)$-particle subspace. Since eigenstates corresponding to systems with different numbers of particles are orthogonal, the state $|n\rangle$ must be an eigenstate of the $N-1$-particle system for the weighting factor to be non-zero. Therefore, $|\langle n|c_{\mathbf{k}}|m\rangle|^2$ is the probability that, upon removal of a particle from the system initially in the eigenstate $|m\rangle$ (in the N-particle system), the system is left in the eigenstate $|n\rangle$ (in the $N-1$-particle system). The energy of the extracted particle is then $\omega = E_m - E_n$. The thermodynamic factor $e^{-\beta E_m}/Z$ is the probability for the state $|m\rangle$ to be initially occupied.

The term $A^-(\mathbf{k}, \omega)$ in the spectral function is therefore associated with particle *removal*, while the $A^+(\mathbf{k}, \omega)$ can be similarly seen to represent particle *injection*. Hence, we intuitively expect the signal from photoemission to be proportional to the term $A^-(\mathbf{k}, \omega)$. Indeed, it can be easily shown that $A^-(\mathbf{k}, \omega) = f(\omega)A(\mathbf{k}, \omega)$, which is essentially the ARPES intensity given by Eq. 3.3.

3.5 Alternative Ways of Viewing Photoemission Data

In Sects. 3.1 and 3.4 three different ways of viewing the photoemission signal were considered; one was the dispersion relation $\varepsilon_{\mathbf{k}}$, the other two related electron counts (intensity) to binding energy (EDC) or to momentum along a specific direction in the BZ (MDC). The latter forms of visualization can be useful in tracking down dispersion bands and their evolution with energy or temperature, which may not be easily done by simply looking at the dispersion map as a whole.

To summarize, the ARPES signal is a function (intensity) of three variables (k_x and k_y momenta and binding energy). The different ways of rendering the data correspond to picking subsets of the above quantities and plotting them against each other. Thus, the dispersion map $\varepsilon_{\mathbf{k}}$ is a plot of the intensity as a function of the binding energy and, say, k_x. EDCs are plots of intensity versus binding energy, and MDCs are plots of intensity versus momentum. One last form of visualization consists in plotting the intensity as a function of both perpendicular momenta at fixed energy. We call this kind of plot a *Constant Energy Surface* (CES). A particularly important CES is the one where the energy is fixed at $E_B = 0$, for it shows the FS.

All these different options for displaying the signal are summarized in Fig. 3.3, which shows the full four-dimensional data set, pictured as a 3-D array containing different values (or colors, which represent the fourth dimension, intensity). The array has as its height the binding energy, while its base consists of the perpendicular

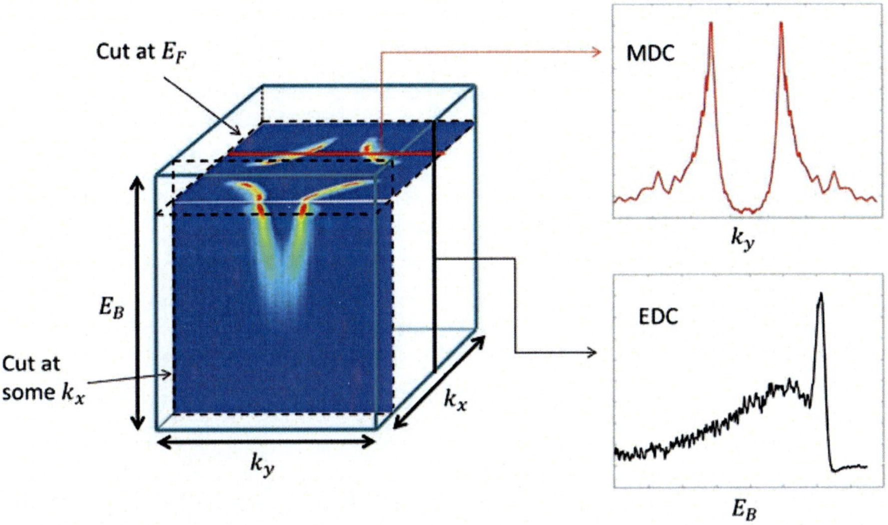

Fig. 3.3 Illustration of different views of ARPES data. Two cross-sections are shown: one parallel to the plane of the page corresponding to the dispersion map $\varepsilon_{\mathbf{k}}$ and another parallel to the base plane corresponding to the Fermi surface. Also shown are two vectors, one parallel to the energy axis representing an EDC taken roughly at the $(\pi, 0)$ point of the BZ (black), the other contained in the plane of the Fermi surface cross-section representing an MDC (red). These particular data correspond to an optimally doped ($T_c = 80$ K) Bi2212 thin film at 20 K

momenta k_x and k_y. Obtaining the different representations of the data amounts to either taking cross-sections of the array volume (for the dispersion maps and the CESs) or picking out vectors of the array (EDCs for vectors parallel to the height of the array and MDCs for vectors parallel to the base plane).

References

1. S. Hufner, *Photoelectron Spectroscopy: Principles and Applications* (Springer Science & Business Media, Berlin, 2013)
2. N.W. Ashcroft, N.D. Mermin, *Solid State Physics* (Brooks/Cole, Pacific Grove, 1976)
3. M.R. Norman, M. Randeria, H. Ding, J.C. Campuzano, Photoelectron escape depth and inelastic secondaries in high-temperature superconductors. Phys. Rev. B **59**(17), 11191–11192 (1999)
4. A. Einstein, Über einen die Erzeugung und Verwandlung des Lichtes betreffenden heuristischen Gesichtspunkt. Annalen der Physik **322**(6), 132–148 (1905)
5. R.L. Withers, J.G. Thompson, L.R. Wallenberg, J.D. FitzGerald, J.S. Anderson, B.G. Hyde, A transmission electron microscope and group theoretical study of the new Bi-based high-T_c superconductors and some closely related Aurivillius phases. J. Phys. C: Solid State Phys. **21**(36), 6067–6083 (1988)
6. G.D. Mahan, Theory of photoemission in simple metals. Phys. Rev. B **2**(11), 4334–4350 (1970)

7. W.L. Schaich, N.W. Ashcroft, Model calculations in the theory of photoemission. Phys. Rev. B **3**(8), 2452–2465 (1971)
8. J.B. Pendry, Theory of photoemission. Surf. Sci. **57**(2), 679–705 (1976)
9. L. Hedin, S. Lundqvist, Effects of electron–electron and electron–phonon interactions on the one-electron states of solids, in *Solid State Physics*, vol. 23 (Academic Press, New York, 1970), pp. 1–181
10. M. Randeria, G. Jennings, H. Ding, J. C. Campuzano, A. Bellman, T. Yokoya, T. Takahashi, H. Katayama-Yoshida, T. Mochiku, K. Kadowaki, Momentum distribution sum rule for angle-resolved photoemission. Phys. Rev. Lett. **74**(24), 4951–4954 (1995)

Chapter 4
Experimental Details

Abstract Experiments in angle-resolved photoemission spectroscopy (ARPES) pose various challenges, like the requirement of ultra-high vacuum, finding a suitable excitation light source, sample preparation and manipulation in vacuum, and high energy and momentum resolution. This chapter contains technical details on the design and construction of the measurement apparatus used in this work. The performance of the instrument in terms of sample surface purity is highlighted.

ARPES is a challenging experimental technique, not only in the sense of gathering data and analyzing them, but also down to the more technical aspects of the design and construction of the measuring instrument. For this reason, it is worthwhile describing the experimental apparatus used in this work in some detail.

In Sects. 4.1 and 4.2 we present an overview of the measurement apparatus and outline the conditioning procedure required for its optimum performance. Sections 4.3 and 4.4 describe the sample manipulator and the ultraviolet (UV) discharge lamp used in this work as a light source. Section 4.5 covers the sample preparation procedure and the novel cleaving technique used in this work. To close this chapter, the performance of the instrument in terms of photon flux, energy resolution, and surface contamination is discussed in Sect. 4.6.

(A detailed account of the instrument, including pumps, pressure gauges, cryostat, electronic wiring, mechanical supports, etc. was deemed superfluous, as most of these components are commonly used in the field and are described in user manuals. Instead, the following description emphasizes the author's contributions and modifications to standard practices. Only the UV discharge lamp is described in some detail.)

F. Restrepo, *Angle-Resolved Photoemission Spectroscopy Study of Spin Fluctuations in the Cuprate Superconductors*, Springer Theses,
https://doi.org/10.1007/978-3-031-10979-9_4

4.1 System Overview

Figure 4.1 shows the principal components of the system: the main vacuum chamber and analyzer (panel a) and the load-lock and transfer chambers (panel b). Although we have partitioned the sketch for clarity, the main and transfer chambers are actually bolted together through the indicated coupling ports, so that the load-lock, transfer, and main chambers are connected in series. After samples are prepared and mounted on appropriate sample holders, they are loaded into the load-lock, which can hold up to four samples. After pumping this volume down to a pressure of $\sim 4 \times 10^{-9}$ Torr, the samples are transferred to the transfer chamber, which is typically kept at a pressure of $\sim 3 \times 10^{-10}$ Torr. The lower transfer arm in this chamber carries a magazine capable of holding up to six samples, although usually only three slots are used for samples and one for the "cleaver" (see Sect. 4.5). Finally, the samples can be transferred to the main chamber, where the photoemission experiments take place. The UV discharge lamp is not visible because it is connected to a port located behind the main chamber in Fig. 4.1a. The sample motion control system (see below) as well as the electron analyzer are connected directly to the main chamber and share the same vacuum with it. This part of the instrument is kept at a pressure below 2×10^{-11} Torr during measurements. The sample magazine in this chamber has two slots, one of which is usually reserved for a gold sample, used as a reference for the chemical potential, as described in Sect. 3.3. This magazine is bolted onto a cold-finger, which runs inside the motion control components all the way up to the cryostat head. Although the closed-cycle, compressed helium cryostat is powerful

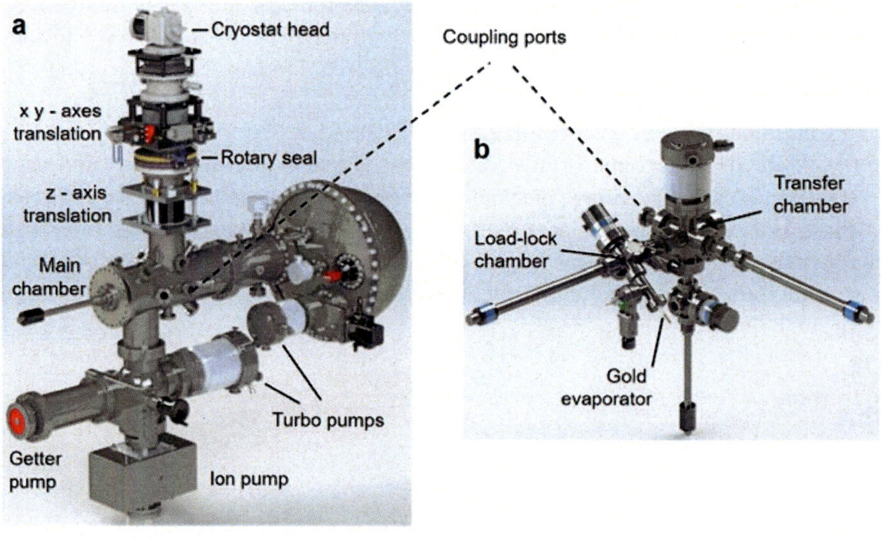

Fig. 4.1 (**a**) Main vacuum chamber and analyzer. (**b**) Load-lock and transfer chambers

enough to cool our samples to $\sim$ 4 K, the Bi2212 single crystal measured in this machine was never cooled below 60 K for this project.

4.2 System Bakeout

As mentioned in Chap. 3, one of the most challenging aspects of photoemission is the requirement of ultra-high vacuum. Electrons have an extremely small escape depth (d) for most solids (for the cuprate Bi2212, $d \sim 3$ Å [1]). That is, only electrons within a distance d of the surface will exit the sample following photo-excitation. If additional contaminants are adsorbed on the surface, electrons will be inelastically and isotropically scattered as they escape the sample, which is deleterious for the angle- and energy-resolving capabilities of an ARPES experiment. As there is no such thing as true vacuum, contamination is inevitable; however, it can be slowed down to a point where experiments can be conducted for days on end without substantial surface contamination. It turns out that the pressures at which this happens are of the order of 10^{-11} Torr. Aggressive pumping of the system is not enough to achieve these pressures in a reasonable time, so that the chamber containing the sample needs to be heated up (or "baked"). Generally, the hotter the chamber gets, the faster its walls desorb contaminants. However, there are temperature-sensitive elements in our system (most importantly, the electron detector) that limit the bake-out temperature to 150 °C.

Many researchers use heating tapes to bake their systems. These long, wrap-around resistors tend to break easily if not handled carefully.[1] Also, heating tapes heat the area of contact much more than other areas, so that heating is not uniform. To circumvent these issues, an oven was built to heat the *air* around the main chamber and analyzer, resulting in a much more even heating process. The oven is simply a box with walls made of fiberglass insulation compressed between two aluminum sheets. Resistive heaters connected in parallel were placed at the bottom of the box, to allow the hot air to rise, and fans were installed at the top of the oven to force the hot air back down and generate convection currents. After several modifications, a design was arrived at which was easy to assemble and yielded pressures below 2×10^{-11} Torr. Figure 4.2 shows pictures of the system with and without the oven. The load-lock and transfer chambers were baked with heating tapes, due to their smaller size and the less stringent vacuum requirements for these spaces.

[1] We learned a painful lesson when one of the tape wires became slightly exposed and touched one of our vacuum roughing pipes, which resulted in an electrical arc and a sizable hole in the pipe. The system was vented and had to be baked again.

a b

Fig. 4.2 Pictures of measurement chamber with and without oven. (**a**) Without the oven one can see the z and $x - y$ axes manipulators, the rotary seal, and part of the cryostat, stacked on top of each other to form the towering structure in the center of the picture. (**b**) The oven covers only the main chamber; smaller chambers outside the oven are the load-lock chamber in the foreground (where samples are loaded) and the transfer chamber further in the back (where the samples are stored before being transferred into the main chamber for measurement). The three chambers are connected in series and in order of decreasing pressure: $P_{\text{load-lock}} \sim 4 \times 10^{-9}$ Torr, $P_{\text{transfer chamber}} \sim 3 \times 10^{-10}$ Torr, $P_{\text{main chamber}} < 2 \times 10^{-11}$ Torr

4.3 Sample Manipulator

Ideally, the sample inside the chamber should have six automated degrees of freedom for its motion: three translational directions and three rotation axes.[2] As this was both technically and economically unfeasible, two rotational degrees of freedom were not automated, but could be varied manually. The automated angle was the azimuthal θ shown in Fig. 3.1b and mentioned in Sect. 3.2. Figure 4.3 shows the manipulator used in this work. z-axis motion (Fig. 4.3a) was accomplished by the home-made linear translation assembly located at the bottom. A vertical bellows was clamped on both ends by aluminum plates, the bottom of which was static, and the top of which was driven up or down by threaded rods. Common manipulators available in the market tend to have either one, two (for smaller bellows) or a greater even number of support rods (for larger bellows). The concept for our manipulator

[2] In this section, the coordinate system is such that the $x - y$ plane is parallel to the ground and the positive z axis points up. The azimuthal angle in the $x - y$ plane is denoted θ.

Fig. 4.3 Motion control unit in our system. (**a**) Vertical (z) motion manipulator. The motor rotates a 2 in. diameter gear (not visible), which drives a 4 in. diameter gear with a belt for mechanical advantage. This larger gear engages the chain drive close to the bottom aluminum plate, making the threaded rods rotate and move the top aluminum plate, and with it the top part of the chamber. (**b**) Rotation (bottom) and $x - y$ (middle) manipulation stages

was motivated by the need to provide z-axis motion while still allowing some rotational flexibility. Using *three* rods allowed for a small additional rotation in the axes which could not be automated. Once the desired tilt of the whole manipulator was chosen, the three rods, which had sprockets at their bases, were joined by a chain, so that all three rods rotated simultaneously. An additional idler sprocket was installed for adjustment of the chain tension. Finally, with the help of Rich Djoutrek at the machine shop at UIC, we designed a fixture by means of which a 6 Amp motor could engage the chain and drive the whole system, which was otherwise a rather exhausting task to accomplish by hand.

Shown in Fig. 4.3b are the rotary seal and $x - y$ stage. These two items were purchased, but their motorization was customized to our needs. Because the system was obviously too home-built to be completely user-proof, it was decided that *full* automation -where a user can control every single motor remotely- was potentially dangerous. Instead, a triple-axis joystick was wired to control three stepper motors corresponding to the x, y, and θ motions, and an additional joystick was wired to control the z-axis motor. This way, the user could walk around and look inside the main chamber while controlling all motion with two joysticks. This arrangement was found to be safe and very convenient.

4.4 The Helium Lamp

By definition, a *photoemission* spectroscopy measurement requires a suitable light source. What constitutes a "suitable" light source is determined largely by context. In this work, since superconductivity in the cuprates is a low energy phenomenon (in the sense that it involves electrons within 100 meV of the Fermi level or less), high photon energy is not necessary, as one is mainly concerned with photoemission from the bonding band. In principle, photon energies slightly in excess of the material's work function (∼4.5 eV) would suffice; in practice, however, most spherical analyzers are not optimized to detect such low-energy electrons and it is best to use higher photon energies. A convenient working range typically found in the literature is $E_\gamma = 10$–40 eV. In this work, a helium lamp capable of delivering 21.2 eV photons was employed. On the other hand, to resolve spectral details at such low energy scales, a very high resolution is imperative. The archetypal example in the context of the cuprate Bi2122 is the superconducting gap, which has the comparatively large value of 40 meV near the $(\pi, 0)$ point in the BZ (at optimal doping). This means that resolutions better than 20 meV are desirable in our experiment.

Ultraviolet discharge lamps (or UV lamps, for short) are capable of delivering continuous, high-flux, meV resolution radiation in the UV range. Moreover, they are easily operated and can be integrated into any home-made system, thereby eliminating the need for large synchrotron facilities. More importantly, in laser-based and synchrotron-based systems, the photon flux tends to fluctuate or even decay with time. This is not a problem in measurements of the *energies* of spectral features (like a peak or a gap), since these energies are independent of intensity; however, a fluctuating photon yield is undesirable in measurements of *spectral intensity*, for the spectra are averaged over extended time periods, during which the flux may vary. In contrast, the photon flux can be easily controlled in an UV lamp, as it is proportional to the helium gas pressure in the discharge chamber. For these reasons, UV lamps are convenient light sources commonly used in the field.

An UV lamp generates UV radiation as a by-product of a well-known physical process, the glow discharge. The next two subsections describe how glow discharges work, with an emphasis on how they are realized in the instrument used in this work (SPECS UVS 300). Then, the nature of the emitted radiation is briefly discussed for the case of helium as a generating medium. The last two subsections deal with the method for plasma confinement in the UVS-300 model—the *duoplasmatron*—and some modifications done on the lamp which enhanced its performance.

4.4.1 Principles of Operation

At the heart of gas discharge lamps lies the more general concept of *plasma discharge*. In the barest terms, one can imagine a gas species being supplied with some energy (typically electrical), as a product of which it either undergoes excitation-relaxation processes—which result in the emission of photons—or ionization, which results in a plasma of electrons, ions, and neutral atoms. These two requirements still leave an ample parameter space, which has given rise to a wide variety of lamp designs [2] differing in their choice of the following parameters:

- Gas species: gases with higher ionization energies yield lines with higher energy.

- Gas pressure: higher pressures mean shorter mean free paths, which in turn promote more complete thermalization of the plasma. Moreover, pressure adjustment enables some control over relative line intensities (see below).
- Electromagnetic field structure: most discharges involve some sort of confinement and acceleration of charged species to maximize the photon yield. There are a number of ways in which this can be accomplished, most of them electromagnetic in nature.
- Discharge volume: smaller volumes allow for smaller gas quantities, which is beneficial if the light source is to be directly coupled to a vacuum system, as is the case in this work.
- Temporal behavior: some plasmas may be excited continuously (d.c. plasmas, as in the present work), whereas some others benefit from more complex temporal structures in the excitation process.

The UV lamp used in this work (SPECS UVS 300) operates with a Direct Current (DC), glow discharge, as described below.

4.4.2 Glow Discharge in the UVS 300 Model

A cathode at ground potential and an anode at a positive potential drive the *electrical discharge*, which in this work is defined as the establishment of an electrical current through a gaseous medium. This current is carried by electrons moving in the cathode-anode direction and by positive ions moving in the opposite direction, owing to the potential difference. Electrons can be introduced into the gas by emission from the cathode. This emission, in turn, may be stimulated by radiation, thermal energy, or electron and atom bombardment. The UVS 300 starts the current with *thermionic emission*, where the cathode is actually a current-carrying filament which emits electrons. The electrons are accelerated towards the anode, in the process of which they may collide with atoms in the gas. The cathode-anode distance, the voltage drop, and the gas pressure are chosen so as to ensure the electrons have enough energy to either *ionize* or at least *excite* the atoms they encounter. Excitation (and subsequent relaxation) give rise to the desired radiation, which is the ultimate purpose of a discharge lamp. Ionization, on the other hand, amplifies the current. Indeed, when an electron ionizes a neutral atom, there remain three particles: a positive ion, the original electron, and the ejected electron. These two electrons may be further accelerated enough to ionize two more atoms, as a result of which there are now four electrons and three ions (counting the one created in the first collision). Clearly, repeated iteration of this process leads to an electron *cascade* and current amplification. However, ionization does not sustain the current; if the external electron pump (filament current in the UVS 300 model) is removed, the discharge will end. This type of discharge is called *dark* or Townsend, and it is weak, in the sense that only a small current (typically in the μA range) flows through the electrodes and consequently very little light output is obtained. Such a weak light source is not suitable for an ARPES experiment.

One must, however, take into account the positive ions that remain after the ionizing collisions. These will be accelerated toward the cathode, where they may knock out *secondary* electrons.[3] If the voltage drop is allowed to keep increasing, more ionization events will take place and more positive ions will reach the cathode and create more secondary electrons. This process will continue until there is a secondary electron being emitted from the cathode for every electron reaching the anode [4], independent of the external pump. The discharge essentially starts to create its own electrons and is called a *self-sustained* or *glow* discharge. The transition from Townsend to glow discharges happens at the *spark* voltage V_S.

Glow discharges have been extensively studied since their discovery and will not be discussed here in detail.[4] Only a few points are worth noting, for they are directly relevant to the present work. First, in a glow discharge there is significantly more ionization than in a Townsend discharge, which means it can be sustained with a much lower voltage. Before entering the glow regime, the UVS 300 lamp starts out with a negligible anode current and an anode voltage of 150V. Once the glow discharge begins, the voltage drops to about 60 V and the anode current rises to 1–3 Amp (compared with $\sim 10^{-6}$ Amp in the Townsend discharge). Second, this comparatively large current leads to more excitation-relaxation events and, hence, to an enhanced light output. Third, there is a minimum pressure for the discharge to be sustained, for an insufficient number of atoms will not produce enough ions and therefore less secondary electrons, resulting in a non-sustained discharge. In this work, where helium gas was used, the minimum operating pressure was $\sim 5 \times 10^{-5}$ Torr. Fourth, once in the glow discharge regime, the current will continue to increase indefinitely and, if no corrective action is taken, an *arc* discharge will take place. The external circuit must be adjusted to keep the current from rising, typically by increasing a variable resistance in series with the anode. The UVS 300 model automatically switches to constant current mode once the glow discharge sets in, keeping the discharge and the photon yield stable.

4.4.3 Basic Notions on the Helium Spectrum

As mentioned above, electrons may also excite atoms, which will quickly de-excite and emit radiation. This radiation comes in the form of discrete lines, owing to the quantized energy spectrum of atoms. The energy of the emitted photon is given by the Bohr condition:

$$E_\gamma = h\nu = E_f - E_i, \tag{4.1}$$

[3] Actually, the dominant process is one in which the positive ion first recombines with an electron at the surface, and then the neutral atom ejects an electron from the cathode [3].

[4] A detailed description of both Townsend and glow discharges may be found in [3, Chaps. 7 and 8].

where E_f and E_i are the energies of the final and initial atomic states, respectively. In practice, the vast majority of the atoms are initially in their ground state at room temperature. Therefore, the radiated power is mostly concentrated on emission lines where the final energy corresponds to the ground state $E_f = E_0$. Also, excitations to lowest-lying energy states are more favorable, since electrons need to accumulate less energy to induce them. Hence, for neutral helium (He I), the most intense line should come from the transition $1s2p \rightarrow 1s^2$,[5] which is indeed the strongest observed line in a helium discharge, with an energy 21.22 eV [5]. From now on, this line will be referred to as *the* He I line, and it is the main line used in this work.

It is important to keep in mind that helium also emits a myriad other lines, some of whose intensities are not entirely negligible. These will not seriously contaminate the spectra if they are sufficiently far away in energy from the main line or if they are much less intense. One important example is the line corresponding to the transition $1s3p \rightarrow 1s^2$, which has an energy of 23.09 eV. This line would give rise to spectra which are copies of the main spectrum (from the main line) shifted by 1.87 eV but with much lower intensity. Since most of the interesting spectral features of the cuprates are within 0.2 eV of the Fermi energy, the unwanted spectra will not affect the main ones. However, as will be discussed in Sect. 4.6, these additional lines affect the energy resolution of the system.

Nothing said hitherto precludes the possibility of a neutral atom being ionized and then the ion being excited, with its resulting decay and emission. The ion, in this case He^+ (or He II), is a hydrogen-like ion consisting of an electron orbiting around a doubly charged nucleus. In fact, neglecting relativistic corrections, the He^+ spectrum is identical to the hydrogen one, with $Z = 2$ [6]. Thus, a transition from the first excited state ($n = 2$) to the ground state[6] ($n = 1$) emits a photon with energy:

$$E_\gamma = 13.6\,\text{eV}\,Z^2\left(1 - \frac{1}{2^2}\right) = 40.8\,\text{eV}, \tag{4.2}$$

which is quite close to the accepted value of 40.819 eV, obtained with high precision calculations [5]. This is the most intense line of He II and henceforth will be referred to as *the* He II line. The UVS 300 can also produce a usable He II output, but its intensity is, at best, two orders of magnitude less than that of He I.

The relative intensities of He I and He II may be controlled by adjusting the gas pressure. Indeed, lowering the pressure increases the mean free path of the electrons, so that by the time they collide with an atom, they have had enough time to accumulate a large kinetic energy. Therefore, the atoms will be excited to higher energies and consequently emit more energetic photons (like He II). By the same token, low energy lines (like He I), will be enhanced at higher pressures. In this

[5] Actually, the excited state $1s2s$ has a lower energy than $1s2p$, but the transition $1s2s \rightarrow 1s^2$ violates the selection rule $\Delta l = \pm 1$.

[6] In this case, the transition is $1s \rightarrow 2p$, subject to the same selection rules as for neutral helium.

work, however, only the He I line was used, since the Scienta electron analyzer is optimized to work at energies around 20 eV. The optimum working helium pressure for the lamp was around $\sim 5\times10^{-5}$ Torr, as measured by a Bayard-Alpert Ionization Gauge calibrated for Nitrogen.

4.4.4 The Duoplasmatron

As mentioned above, the photon yield can be optimized by exploiting the freedom in discharge configuration design. In particular, different methods exist for confining the plasma in ways that maximize electron-atom collisions and photon yield. One such confinement method is called *duoplasmatron*, and it is the design upon which the SPECS UVS 300 is built. Interestingly, the concept was originally conceived by M. V. Ardenne as a means to create high flux, directional, *proton* beams. Six years later, the first applications of the principle to the creation of ultraviolet radiation were reported [7]. The aim of the duoplasmatron is to confine the plasma (and hence the emitted photons) in one dimension which, for a cylindrical geometry, is the axial direction.[7] One way to accomplish this is by using magnetic fields; in particular, charged particles will propagate helically around axial field lines, thus eliminating any radial motion and divergence from the axial direction. Additionally, one could pick out a small cross-section of the plasma by a suitable aperture, thereby extracting an axially propagating, small (in the transverse direction) plasma beam. This plasma will then generate photons with similar characteristics through excitation-relaxation. Figure 4.4a shows the basic components of a the duoplasmatron setup used in ref. [7]. It depicts the external magnets, which create the axial magnetic field, as well as a *baffle electrode* which picks out the central portion of the plasma. Note that this electrode is also magnetized, so that there is a strong magnetic field at the tip of this electrode, just right of the anode. This is where the highest plasma density is confined, and where most of the radiation comes from (see also point 3 in Fig. 4.4b). The generated radiation is coupled to a copper capillary (shown on the left of Fig. 4.4b); this is followed by a quartz capillary, which enters the vacuum chamber and ends about 2 cm from the sample (not shown).

4.4.5 Modifications to the UVS 300 Lamp

One drawback of the UVS 300 lamp is the large spot size ($\sim$ 2 mm). Since the measured samples were roughly 1mm^2 in area, a considerable portion of the beam did not hit the sample, so that the *actual* photon flux was lower than expected. This limitation led to a chain of undesirable consequences, starting with longer

[7] For the rest of this section, the geometry is assumed to be cylindrical, since this is the design most manufacturers use.

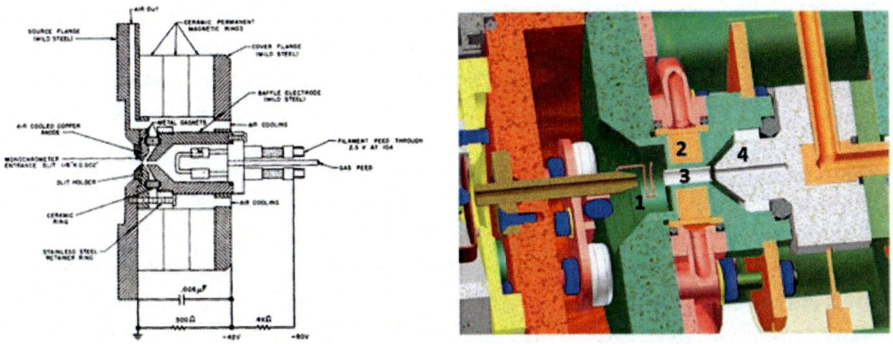

Fig. 4.4 Duoplasmatron configuration. (**a**) Schematic of design used in [7], showing filament (cathode), external magnets, baffle electrode, and anode. (Reproduced from Rev. Sci. Instr. 33(12):1340–1343 (1962), with the permission of AIP Publishing https://doi.org/10.1063/1.1717774). (**b**) CAD drawing of SPECS UVS 300 lamp [8], zoomed in the discharge region. Shown are the filament (1), magnet (2), discharge volume (3), and anode (4) (reproduced from UVS 300 Manual, with the permission from SPECS)

acquisition times. This, in turn, imposed more stringent requirements on the vacuum purity and also forced us to cleave the samples more often, which also meant more samples had to be used up in a shorter amount of time.

An attempt was made to alleviate the problem, consisting in drawing the quartz capillary to a smaller diameter. The original capillary had an inner diameter (ID) of 1 mm throughout its entire length. This was replaced by a capillary having an ID of 2 mm near the discharge zone (in the lamp) and tapered down to 0.5 mm at the other end (near the sample) by mechanically pulling while applying heat. This new capillary was to reduce beam divergence and the spot size on the sample. However, there was considerable scattering from the capillary end and the photon flux was diminished even more. Although the concept could be improved by a more careful tapering procedure and a subsequent frosting of the capillary end surface, time constraints forced us to keep working with the original capillary.

Another modification concerned the type of gas used for the discharge. Commercial UV lamps require gas purities of 99.999% to start a discharge. While this may sound like a very high purity, it actually is not so for ultrahigh vacuum measurements. Indeed, in the first attempts of this experiment, only a commercial, ultra-high purity helium bottle was used, along with thoroughly cleaned stainless steel pipes. The discharge started successfully, but the photoemission signal was virtually indiscernible from the noise. A mass spectrum analysis revealed several contaminants were entering the chamber from the helium lamp and possibly contaminating the sample, which explained the poor signal-to-noise observed. The problem was exacerbated by the fact that the gas entered the chamber through a quartz capillary ending less than 2 cm away from the sample, so that the sample was directly facing an incoming flow of impure gas.

As a first attempt at fixing the problem, a liquid Nitrogen (LN2) trap was used. LN2 cools the pipes enough to condense impurities in the gas line, without condensing helium itself (which condenses at lower temperatures), effectively working as a cryo-pump. This substantially improved the signal-to-noise ratio, but the quality of the data was not high enough to resolve the desired spectral details. Moreover, it would take less than an hour before the sample began to show signs of surface contamination. Since the photon flux was not too high (Sect. 4.6), acquisition times exceeded this time window, especially for full BZ scans.

The next step in improvement came in the form of a very high purity gas microfilter, capable of delivering helium with a purity in the "parts per trillion" range (SAES - MC1 902). This was a significant improvement over any filtering mechanism we devised. The filter was installed after the LN2 trap, right before the valve to the UV lamp. Due to the very pure vacuum in the system and the increased purity of the helium gas, the samples could then be measured for up to one week without signs of surface contamination, at which point they could simply be cleaved again and continue the experiment. The filter, however, is highly sensitive to atmospheric contamination, showing a marked decrease in performance as a function of exposure time. After about nine months of intensive use, the filter began to show signs of degradation, as evidenced by the presence of CO gas in the mass spectrum. Therefore, a second filter was purchased and installed in series with the first one. Also, the piping leading from the helium gas bottle to the lamp was replaced with VCR piping -where the joints are sealed with stainless steel gaskets- and was thoroughly checked to be leak-tight. These modifications led to a substantial improvement in system performance, as detailed in Sect. 4.6.3.

4.5 Sample Preparation

As mentioned in Chap. 2, the cuprate Bi2212 consists of bilayers of copper-oxygen planes weakly coupled via van der Waals forces. This is important mainly because ARPES measurements are more easily interpreted in 2-D samples (Sect. 3.3). But, from a more practical point of view, this two-dimensionality also greatly facilitates the cleaving process, whereby the upper (contaminated) layer of the sample is mechanically removed in vacuum, leaving a fresh new sample surface. The traditional cleaving method used by many groups has been to glue an aluminum pin on the sample surface, inserting the sample/pin system into the vacuum chamber, and removing the pin with the aid of some mechanical feedthrough. One drawback of this method is that, once the sample inevitably becomes contaminated (or if it does not cleave properly), it has to be removed from the main chamber and taken out into the atmosphere, where the pin must be replaced for a new cleave. This whole process takes several hours, since the load-lock chamber (see Fig. 4.1b) needs to be baked before the sample can be transferred back into the main chamber.

It is more desirable to have a cleaving method where the sample can be cleaved successively without having to take it outside the vacuum. To this end, the bottom

Fig. 4.5 “Cleaver” made by covering bottom of a copper slug with double-sided polyimide tape. Several Bi2212 layers can be seen as the result of successive sample cleaving. Near the bottom a hole can be seen where one of the pins of the mechanical feedthrough can engage and transfer the cleaver

face of one of the sample holders was covered with double-sided polyimide tape, which is compatible with UHV, as shown in Fig. 4.5. This new “cleaver” could be easily introduced into the load-lock chamber and kept in the transfer chamber until it was needed. When the time came, one simply had to pick it up with one of the mechanical feedthroughs and bring it up to the sample until it made contact. The adhesive from the polyimide tape then cleaved the top layer of the sample and the cleaver could then be retracted and stored outside the main chamber. This process could be repeated as needed with the same sample until it became too thin.[8] Optimally doped and overdoped Bi2212 single crystals could be successively cleaved up to five times, depending on their original thickness, with the samples never leaving the main chamber.

[8] The last cleave usually resulted in a broken sample, with some of the underlying glue exposed. The bad quality of the cleave was immediately apparent from the distorted spectra collected by the analyzer.

4.6 System Characterization

In the next two subsections, data from two different Bi2212 samples are compared. The first data set corresponds to an optimally doped Bi2212 thin film, collected by former group member Utpal Chatterjee at the Synchrotron Radiation Center in Madison, Wisconsin. The second data set corresponds to an optimally doped Bi2212 single crystal, collected by the author and his colleague, Hao Xu, with a helium discharge lamp at the University of Illinois at Chicago. The specifications for the helium lamp all pertain to the He I line emitted from the SPECS UVS 300 lamp.

4.6.1 Flux

The *nominal* flux of the He I line (21.22 eV) is of the order of 10^{12} photons/sec, comparable to synchrotron radiation [8–10]. However, since our beam diameter was roughly 2 mm, while our single crystals were about 1 mm, the actual flux impinging on our sample was substantially lower than the nominal flux of the lamp. To estimate the lamp flux, we first took a spectrum of a large ($\sim$ 3 mm diameter) polycrystalline gold sample, to capture the entire beam, and compared it with the corresponding spectrum collected with synchrotron radiation. After normalizing for total acquisition times and bin sizes, we found that the spectra had almost equal intensities, so the total output of the helium lamp is quite similar to that of a typical synchrotron source -about 10^{12} photons/second. To estimate the photon flux actually impinging on our samples, we turned to smaller Bi2212 single crystals and performed an analogous comparison with Bi2212 thin films (at a similar temperature and at the same point in the BZ, to minimize differences in matrix element effects). In this case, we saw that the synchrotron spectra, for which the entire beam hit the sample, had an intensity about 4 times larger than the UV-lamp data. Hence, we concluded that the actual flux *on the sample* was in the 10^{11} photons/sec regime. This reduced flux implied much longer measurement times, which is undesirable in view of the inevitable contamination of Bi2212 surfaces. This difficulty was the main motivation behind the new cleaver design and the helium purification efforts described in Sects. 4.4.5 and 4.5.

4.6.2 Energy Resolution

The energy resolution of the He I line is $\sim$ 1 meV. Convolved with an analyzer resolution of $\sim$ 1 meV, the overall energy resolution of the system should be $\sim$ 1 meV. The actual energy resolution was obtained by fitting a resolution-broadened Fermi function to a polycrystalline gold spectrum at 20 K, as shown in Fig. 4.6.

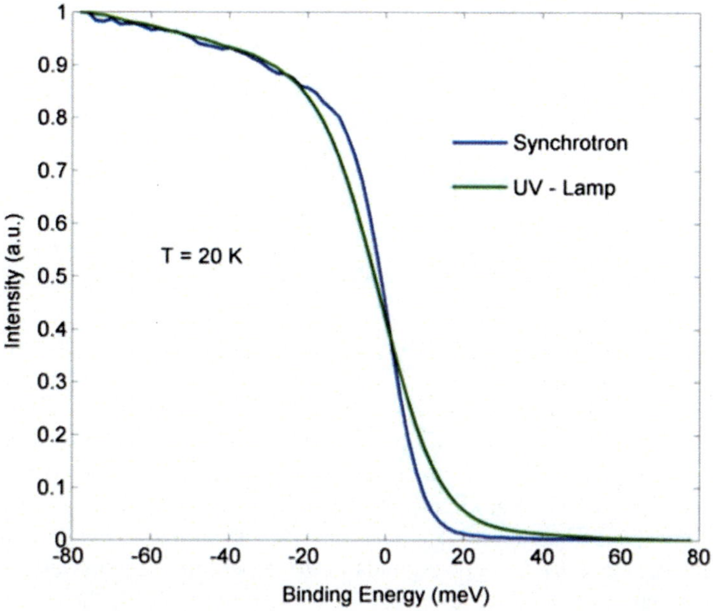

Fig. 4.6 Comparison of Au spectra taken at the Synchrotron Radiation Center (Madison, Wisconsin) with 22.1 eV photons (blue) and with the He-I line from an UV - lamp (green). Spectra were collected at 20 K and normalized to unity at −80 meV

The fitted curve was of the form

$$\int d\omega' R(\omega - \omega') \frac{a + b\omega'}{e^{(\omega' - \mu)\beta} + 1}, \tag{4.3}$$

where $R(w) = \frac{1}{\sigma\sqrt{2\pi}} \exp\left[- \left(\frac{\omega}{\sqrt{2}\sigma}\right)^2 \right]$ is the resolution function, μ is the chemical potential, $\beta = 1/kT$, and the constants a and b are used to model the downward slope of the spectrum, product of secondary background photoemission. The full width at half maximum (FWHM) of $R(\omega)$ ($\sim 2.3\sigma$) was taken as a measure of the energy resolution. The result of the fit was FWHM $\sim$ 25 meV, which is substantially larger than the width of the synchrotron spectrum shown in Fig. 4.6, $\sim$ 15 meV. The reason the resolution is not as low as estimated is most likely the presence of other He lines in the lamp. Usually these lamps are used with monochromators or filters to select only the desired He-I line. Without these, one gets photoemission from *all* helium lines. As mentioned in the beginning of Sect. 4.4, a resolution of 25 meV is still quite sufficient to resolve the most important low-energy features of HTSCs.

4.6.3 Sample "Aging"

Although not the main subject of this work, in this section we digress on some results which are interesting from a purely experimental point of view and having to do with extending the sample lifetime in vacuum. Typically, ARPES measurements tend to be limited by surface contamination, implying the need to cleave samples every several days until they are used completely. Here, we describe how we managed to measure a sample for about six weeks *after a single cleave*. To this end, we compare the antinodal ARPES spectra for an optimally doped Bi2212 single crystal at $T = 60$ K at different times after a single cleave.

Figure 4.7a shows the dispersion along the BZ edge (in the direction $\bar{M} - Y$) for the sample the day it was cleaved. A gap, followed by a substantial enhancement of the density of states near the Fermi energy ($\omega = 0$) is observed—consistent with the formation of Bogoliubov quasiparticles [11]—along with the well documented superlattice signal. The three horizontal lines (labeled 1, 2, and 3) represent the locations of the EDCs that were plotted in Fig. 4.7d–f. EDCs 1 and 2 correspond to two non-equivalent antinodal positions (the symmetry being broken by the superlattice), while EDC 3 corresponds to the superlattice exclusively. The lack of surface contamination is already apparent in Fig. 4.7b, showing the same dispersion as panel (a), but after 25 days of continuous measurements. A more detailed comparison of the selected EDCs is shown in panels (d)-(i); in particular, the reduced energy scale in panels (g)-(i) shows only minute variations (order 5 meV, well below our resolution) in the peaks and leading edges of the spectra. More strikingly, the data after 42 days (without cleaving) still show little sign of sample aging (see panel *c* and green curve in all other panels). We stress that, aside from a subtraction of the lowest intensity from each EDC, all data presented in Fig. 4.7 are raw and were not even normalized at any energy. The data were collected with the exact same detector parameters, the same acquisition times, and with the same UV lamp flux, which was fixed by keeping the He pressure in the lamp constant.

Even though the pressure in the chamber was below our detection limit of 2×10^{-11} Torr and the double filter arrangement ensured a helium influx of extremely high purity, we suspect that these improvements are not enough to explain such a complete lack of sample aging. In their ARPES experiments on Bi2212, Palczewski et al. [12] reported that, even at pressures similar to ours, there was still a minute concentration of CO and CO_2 contaminants in the measurement chamber. Furthermore, they noted that these contaminants could deposit on the surface and become dissociated due to UV radiation, allowing the remaining oxygen atoms to diffuse into the sample and effectively dope it. Indeed, the authors saw the emergence of an antibonding band, indicative of sample overdoping. Finally, they claimed that heating the sample to 280 K could effectively remove the oxygen and bring it back to its original doping level. As is clear from Fig. 4.7, we saw no sign of sample doping. However, after a full day of measurements, we did observe a slight broadening and shift ($\sim$ 4 meV) to lower kinetic energies of spectral features. We believe this was caused by the helium gas adsorbing on the sample surface, slightly

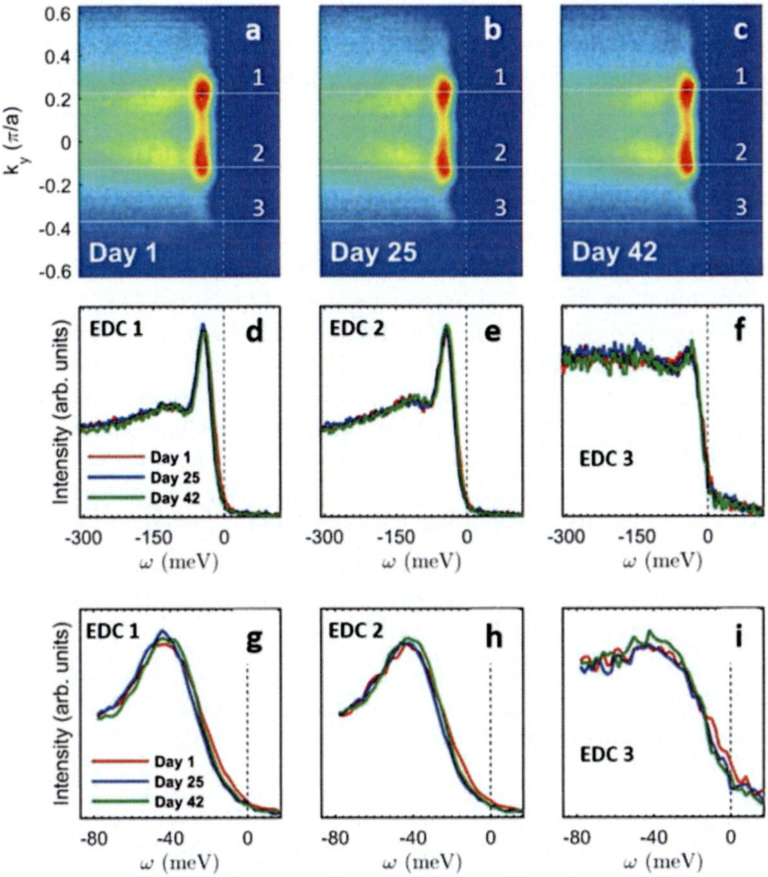

Fig. 4.7 Absence of contamination in superconducting Bi2212 as a function of time. (**a**)–(**c**) Antinodal dispersion along $\bar{M} - Y$ direction. (**d**)–(**f**) Selected EDCs (indicated in panels (**a**)–(**c**) for different days. (**g**)–(**i**) Same as (**d**)–(**f**), but in a reduced energy scale. Vertical scales in (**d**)–(**i**) are not the same; they were adjusted to fit the spectra. Redundant energy or momentum scales in a given panel should be read off adjacent panels (reprinted from Restrepo et al. Coupled electronic and magnetic excitations in the cuprates and their role in the superconducting transition. Commun. Phys. **5**, 45 (2022))

"slowing" down the photoemitted electrons. Indeed, the spectra collected at day 25 (blue curves in Fig. 4.7d–i) were collected before this issue was realized, and the peaks show a slight drift to lower kinetic energy. Once this was realized, the sample was heated every day up to at least 250 K before starting new measurements.

To explain the slow surface contamination, we speculate that a thin layer of adsorbed helium atoms on the surface protected it from contaminants like H_2O, CO_2, and CO. Since the sample was continuously measured at temperatures below 150 K, and not allowed to heat up overnight, the surface was kept at cryogenic temperatures most of the time. We only heated the sample quickly before starting the measurement, so as to remove adsorbed helium, but then we cooled it down immediately for measurements. In contrast to reference [12], we did not heat our sample to alter doping, but rather simply to release the weakly bound helium atoms from its surface. Incidentally, we never got to see this sample actually show signs of contamination, since our experiment was interrupted by technical complications at Day 44 after cleaving the sample.

The above explanation is hypothetical at this stage and more controlled experiments need to be conducted to establish the mechanism behind such slow surface contamination. However, we believe our observations are of use in research requiring ultrahigh vacuum at low temperatures (ARPES, STM, surface science, among others). In particular, for ARPES, we note that this method of extending sample lifetime with helium adsorption is not exclusive to helium lamp systems, as any inert gas can be easily introduced into any vacuum system where the sample is cold enough to allow condensation. The only caveat is that the residual contaminants in the gas have to be brought down to negligible concentrations (in the parts per trillion range), to avoid contaminating the sample with the gas itself. As noted in Sect. 4.4.5, this can be accomplished with ultra-high purity microfilters

References

1. M.R. Norman, M. Randeria, H. Ding, J.C. Campuzano, Photoelectron escape depth and inelastic secondaries in high-temperature superconductors. Phys. Rev. B **59**(17), 11191–11192 (1999)
2. A. Bogaerts, E. Neyts, R. Gijbels, J. van der Mullen, Gas discharge plasmas and their applications. Spectrochim. Acta Part B: Atom. Spectrosc. **57**(4), 609–658 (2002)
3. J.D. Cobine, *Gaseous Conductors: Theory and Engineering Applications*, 1st edn. (McGraw-Hill, New York, 1941)
4. J.H. Beynon, *The Conduction of Electricity Through Gases* (Harrap, London, 1972)
5. J.E. Sansonetti, W.C. Martin, Handbook of basic atomic spectroscopic data. J. Phys. Chem. Ref. Data **34**(4), 1559–2259 (2005)
6. H.A. Bethe, E.E. Salpeter, *Quantum Mechanics of One- and Two-Electron Atoms* (Springer, Berlin, Heidelberg, 1957)
7. J.A.R. Samson, H. Liebl, Duoplasmatron as a vacuum ultraviolet light source. Rev. Sci. Instrum. **33**(12), 1340–1343 (1962)
8. *UVS 300 Very High Intensity UV Lamp and UVS 300 - A Power Supply Manual Version 1.2* (SPECS GmbH, Voltastraße, Berlin, 2008)
9. M.H. Berntsen, O. Götberg, O. Tjernberg, An experimental setup for high resolution 10.5 eV laser-based angle-resolved photoelectron spectroscopy using a time-of-flight electron analyzer. Rev. Sci. Instrum. **82**(9), 095113 (2011)
10. Y. He, I.M. Vishik, M. Yi, S. Yang, Z. Liu, J.J. Lee, S. Chen, S.N. Rebec, D. Leuenberger, A. Zong, C.M. Jefferson, R.G. Moore, P.S. Kirchmann, A.J. Merriam, Z.X. Shen, High resolution angle resolved photoemission with tabletop 11 eV laser. Rev. Sci. Instrum. **87**(1), 011301 (2016)
11. N.N. Bogoljubov, On a new method in the theory of superconductivity. Il Nuovo Cimento (1955–1965) **7**(6), 794–805 (1958)
12. A.D. Palczewski, T. Kondo, J.S. Wen, G.Z.J. Xu, G. Gu, A. Kaminski, Controlling the carrier concentration of the high-temperature superconductor $Bi_2Sr_2CaCu_2O_{8+\delta}$ in angle-resolved photoemission spectroscopy experiments. Phys. Rev. B **81**(10), 104521 (2010)

Chapter 5
Results

Abstract The formation of Cooper pairs, a bound state of two electrons of opposite spin and momenta by exchange of a phonon, is a defining feature of conventional superconductivity. In the cuprate high temperature superconductors, even though it has been established that the superconducting state also consists of Cooper pairs, the pairing mechanism remains intensely debated. In this chapter, we investigate superconducting pairing in the $Bi_2Sr_2CaCu_2O_{8+\delta}$ (Bi2212) cuprate by employing spectral functions obtained directly from angle-resolved photoemission (ARPES) experiments as input to the Bethe-Salpeter gap equation. Assuming that Cooper pairing is driven solely by spin fluctuations, we construct the single-loop spin-fluctuation-mediated pairing interaction, and use it to compute the eigenfunctions and eigenvalues of the Bethe-Salpeter equation in the particle-particle channel for multiple Bi2212 samples. We find that, as the temperature is reduced, the leading eigenvalue increases upon approaching T_c, reaching a value of approximately unity at the T_c corresponding to each doping value. Moreover, the Bethe-Salpeter eigenfunctions, which reflect the properties the pairing interaction, have $d_{x^2-y^2}$ symmetry in k-space, as observed experimentally. These results suggest that spin fluctuations are a viable candidate for the pairing "glue" in the cuprate high temperature superconductors.

As mentioned in the introduction and in Chap. 2, the critical temperature for a superconducting transition can in general be determined from a linearized gap equation (BSE)

$$\lambda(T)\Phi(T) = \hat{O}(T)\Phi(T), \tag{5.1}$$

F. Restrepo, *Angle-Resolved Photoemission Spectroscopy Study of Spin Fluctuations in the Cuprate Superconductors*, Springer Theses,
https://doi.org/10.1007/978-3-031-10979-9_5

where the operator $\hat{O}$ is defined through

$$\hat{O}(T)\Phi(\mathbf{k},\omega_n,T) \\ = -\frac{T}{N}\sum_{\mathbf{k}',\Omega_n} V(\mathbf{k}-\mathbf{k}', i\omega_n - i\Omega_n, T)G(\mathbf{k}', i\Omega_n, T)G(-\mathbf{k}', -i\Omega_n, T)\Phi(\mathbf{k}', \Omega_n, T). \tag{5.2}$$

In this work, we assume a form of the pairing interaction mediated by spin excitations which has been hypothesized to describe the unconventional cuprate superconductors [1]:

$$V = V_{\mathrm{SF}}(\mathbf{q}, i\omega_m, T) = \frac{3}{2}U_0(T)^2\chi(\mathbf{q}, i\omega_m, T). \tag{5.3}$$

The Green's function G and the spin susceptibility χ appearing in the BSE uniquely describe the system undergoing the superconducting transition; in our case, this system consists of a collection of interacting electrons confined to a CuO_2 plane of the cuprate Bi2212. For computational efficiency (see Appendix A), these quantities have been written in Matsubara formalism, where the excitation energies for a system of fermions (bosons) take the discrete values $\omega_p = p\pi T$ (p is an odd (even) integer). These are not measurable energies, and the functions $G(\mathbf{k}, i\Omega_n)$ and $\chi(\mathbf{q}, i\omega_m)$ cannot be probed by any experiment. However, these functions are mathematically related to real observables through the *spectral representations* [2]

$$G(\mathbf{k}, i\omega_n) = \int d\omega \frac{A(\mathbf{k},\omega)}{i\omega_n - \omega} \quad \text{and} \quad \chi(\mathbf{q}, i\omega_m) = -\frac{1}{\pi}\int d\Omega \frac{\mathrm{Im}\chi(\mathbf{q},\Omega)}{i\omega_m - \Omega}, \tag{5.4}$$

where $A(\mathbf{k},\omega)$ is the spectral function measured in ARPES (Sect. 3.4) and $\mathrm{Im}\chi(\mathbf{q},\Omega)$ is the imaginary part of the spin susceptibility, measured in INS experiments (Sect. 2.4). For this latter quantity, though, we did not use INS data, but rather constructed it from the spectral function via the commonly employed Random Phase Approximation (RPA), as shown below. This way, all factors entering the BSE were ultimately derived from ARPES data collected from the cuprate Bi2212.[1]

The "true" gap equation $\Phi = \hat{O}\Phi$ (that is, the BSE without the auxiliary eigenvalue) is a generalization of the BCS gap equation (Eq. 2.16) which allows for the pairing interaction and the order parameter to have an energy dependence. Recall that this latter equation was derived under the assumption that the system was close to the transition temperature. This means that the equality in these gap

[1] As we show in this chapter and in Appendix B, the spin-fermion coupling U_0 can be estimated in two independent ways, both entirely dependent on the spectral function as well.

equations *only holds at* T_c. In principle, we could use this fact to assess whether the spin-fermion interaction V_{SF} can reproduce the observed values of T_c in Bi2212. Indeed, we could use Bi2212 data collected at T_c, construct the operator $\hat{O}$, and check whether there is a solution Φ satisfying the true gap equation. If this were the case, we would then say that the pairing interaction V_{SF} can adequately describe a superconducting transition in Bi2212 taking place at the measured T_c. This procedure, however, is too restrictive, for it is highly unlikely that an operator built from experimental data will map a non-trivial function onto itself. For this reason, it is convenient to introduce the auxiliary eigenvalue λ in the BSE. We can then solve this equation using ARPES data taken at various temperatures and track the temperature dependence of the extracted eigenvalues. The temperature at which $\lambda \sim 1$ then represents the value of T_c predicted by the Bethe-Salpeter formalism where the pairing interaction is mediated by spin fluctuations. Our principal result is that, for various doping levels, the approximate T_c values thus obtained are close to the *actual* transition temperatures in Bi2212 and, consequently, that the spin-fermion interaction may be the source of the unconventional superconducting state in the cuprate superconductors.

Throughout the several steps required to arrive at this result, other results were obtained, which also shed light on the doping and temperature dependence of spin fluctuations and their interaction with fermions in the cuprates. This chapter presents the details of the procedure and the results, divided into the following sections:

- Section 5.1. We present some of the ARPES data employed and explain how the spectral functions $A(\mathbf{k}, \omega)$ were extracted from them.
- Section 5.2. We use the experimentally derived spectral functions to construct the bare and interacting spin susceptibilities in real energies. In this section, we also point out an artifact coming from one of the assumptions made in extracting the spectral function, we show how it affects the calculation of the spin response, and suggest a way to mitigate its effects. We then present the calculated spin response and draw some comparisons with the spin susceptibilities measured in INS of Bi2212 and other cuprates. Besides allowing us to cross-check the calculations, enforcing consistency with INS data was necessary to extract the spin-fermion coupling U_0 at different doping levels and temperatures. (In Appendix B, we present an alternative method for extracting this energy scale.) We also comment on the behavior of the spin resonance intensity with temperature and doping, and on the spectral weight distribution in the calculated spin response.
- Section 5.3. For the pairing problem, we switch to Matsubara energy calculations, which are less computationally demanding. The Matsubara Green's function and spin susceptibility are obtained from simple Lehmann representations of the spectral function and the spin response (also obtained from the spectral function) respectively. These quantities, along with the coupling energy U_0, are used to construct the operator $\hat{O}$ in Eq. 1.4 and solve for the Bethe-Salpeter eigenvalue and eigenfunction. Absent experiments in Matsubara energies, we compare the calculated pairing eigenfunctions with results from the Hubbard model calculations of Scalapino and co-workers. Motivated by the consistency of

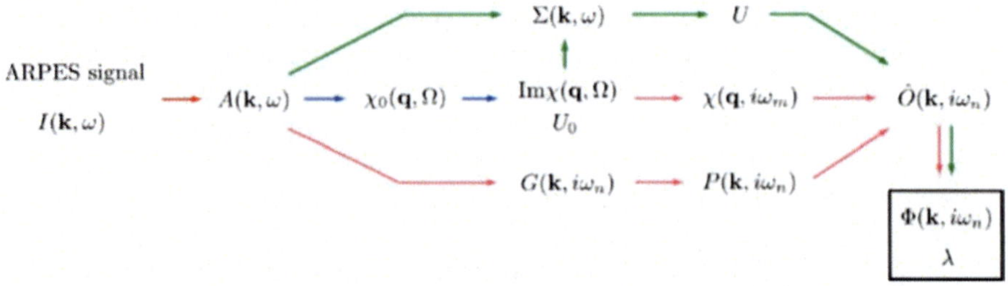

Fig. 5.1 Roadmap of calculations presented in this chapter. Steps are color coded as follows: Red—Sect. 5.1; Blue—Sect. 5.2; Pink—Sect. 5.3; Green—Appendix B (reprinted from Restrepo et al. Coupled electronic and magnetic excitations in the cuprates and their role in the superconducting transition. Commun. Phys. **5**, 45 (2022))

these latter calculations with our phenomenological approach, we finally present our main results for the temperature dependence of the leading eigenvalue λ and the estimated transition temperatures.

For reference, the above steps are summarized in the workflow diagram in Fig. 5.1. The remaining sections in this chapter are devoted to discussions of the uncertainties in the reported eigenvalues (Sect. 5.4) and the obtained spin-fermion couplings (Sect. 5.5).

5.1 ARPES Data and the Spectral Function

As mentioned in Chap. 3, ARPES probes electronic excitations in momentum space as a function of energy. In the all-important CuO_2 planes of the cuprates, the reciprocal lattice is represented by a square BZ with symmetry points labeled as follows: $\mathbf{k} = (0,0) \rightarrow \Gamma$; $\mathbf{k} = (\pi/a, 0) \rightarrow \bar{M}$; $\mathbf{k} = (\pi/a, \pi/a) \rightarrow Y$; $\mathbf{k} = (\pi/a, -\pi/a) \rightarrow X$ (where a is the lattice constant). Figure 5.2a shows a tight-binding fit of the Bi2212 bonding band at the Fermi energy; that is, the Fermi surface. The 15 narrow rectangles show the approximate extent of momentum space sampled in this work. As noted already, the angles θ_i represent discrete sample rotations, while ϕ represents the *multiplexing* angles, which are much more densely spaced. The data have an additional energy dimension going in and out of the plane of the page (Chap. 3).

As indicated in Fig. 5.2b, which is a reproduction of Fig. 1.1 included here for convenience, the ARPES data used in this work were collected from three different Bi2212 samples at two doping levels: an underdoped thin film with $T_c = 67$ K (UD67, black circles), an optimally doped thin film with $T_c = 80$ K (OP80, red diamonds), and an optimally doped single crystal with $T_c = 91$

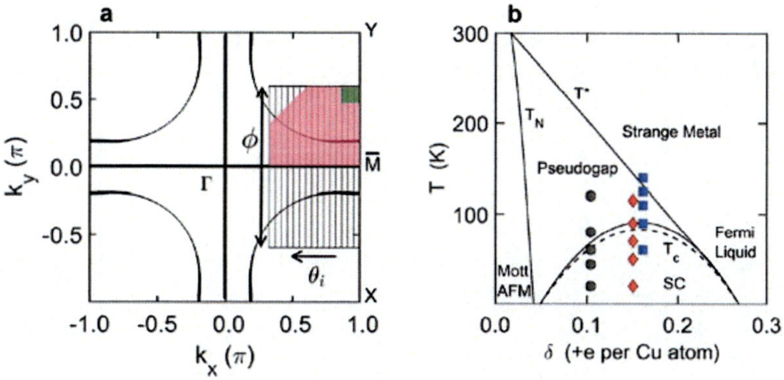

Fig. 5.2 (**a**) Tight-binding Fermi surface of Bi2212 [3]. The narrow rectangles show the approximate region of the BZ covered in the measurements. The pink area is the portion of the data selected for reflection about the symmetry axes of the zone. The green shaded area is the momentum region where the background was extracted and averaged for the thin film samples. (**b**) Simplified phase diagram of Bi2212. Black circles correspond to the UD67 sample, red diamonds correspond to the OP80 sample, and blue squares to the OP91 single crystal. These symbols will be used consistently throughout this chapter. The dashed SC dome applies to UD67 and OP80, while the solid one corresponds to OP91 (adapted from Restrepo et al. Coupled electronic and magnetic excitations in the cuprates and their role in the superconducting transition. Commun. Phys. **5**, 45 (2022))

K (OP91, blue squares).[2] As will be seen in the next two sections, the integrals involved in the calculations of the spin susceptibility and the solution of the BSE require spectral functions measured over an entire BZ and for energies covering the bonding band of Bi2212. Also, since the spin susceptibility constructed from spectral functions is highly sensitive to the momentum anisotropy of the spectral gap in Bi2212, sufficiently high energy resolution is required as well to resolve this feature. Altogether, the ARPES data used in this work amount to 15 points in the phase diagram, covering different doping levels and temperatures, and each point consisting of a comprehensive energy-momentum sampling of the electronic excitations.

Representative raw ARPES data for the OP91 single crystal are presented in Fig. 5.3. The right two panels show the Fermi surfaces measured at 60 K and 125 K, laid over the Fermi surface calculated from a tight-binding fit to the dispersion [4]. The main band (laid over thick yellow lines in the tight-binding fit) and the superlattice bands (laid over thin yellow lines in the tight-binding fit) are visible. For both temperatures, note the stronger signal from the lower branch of the main band, which is simply an enhancement due to superlattice bands overlapping in this region of the BZ. Note also the difference in intensities between the superconducting and normal state data, due to the opening of the superconducting gap, which

[2] The data sets for the UD67 and OP80 thin films were collected by former group member, Utpal Chatterjee. The OP91 data were collected by the author and his colleague, Hao Xu.

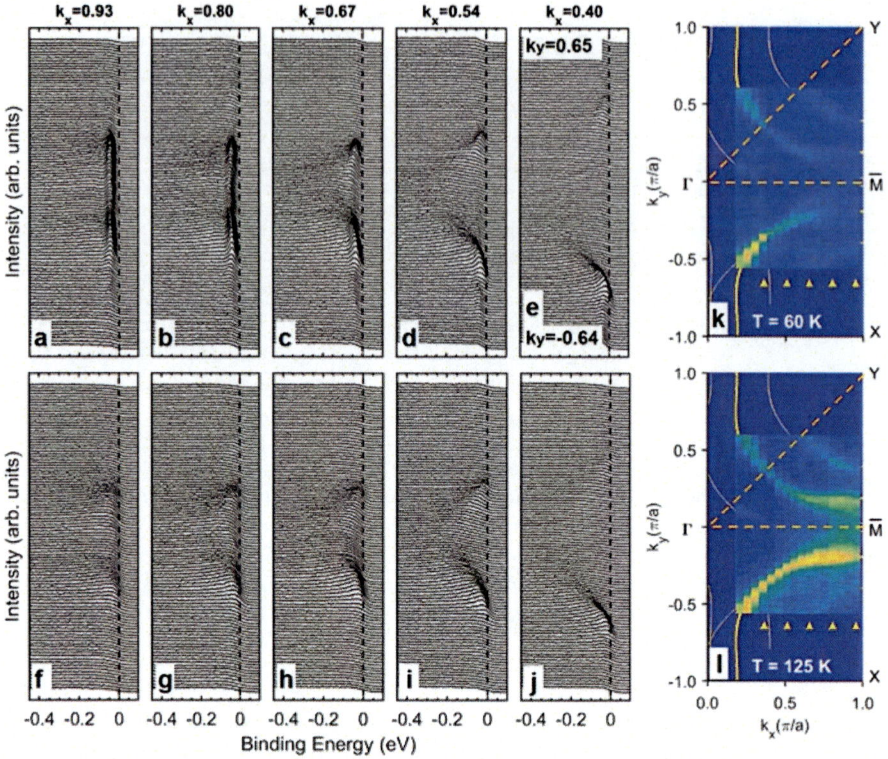

Fig. 5.3 Selection of ARPES data from the Bi2212 single crystal (OP91) in the normal and superconducting states. (**a**) EDCs located at coordinates $\mathbf{k} = (0.97, -0.64 \rightarrow 0.65)\pi/a$ with $T = 60$ K (slice indicated by rightmost triangle in panel k). EDCs in panels (**b**)–(**e**) correspond to progressively lower values of k_x (indicated by triangles in panel (**k**). Panels (**f**)–(**j**) show the same exact EDCs but at $T = 125$ K. (**k**) and (**l**) are the Fermi surfaces measured at $T = 60$ K and $T = 125$ K, respectively, laid over a tight-binding fit to the dispersion. (Adapted from Restrepo et al. Coupled electronic and magnetic excitations in the cuprates and their role in the superconducting transition. Commun. Phys. **5**, 45 (2022))

suppresses electronic states at the Fermi energy. In particular, the intensity in the superconducting state (panel k) is lowest at the antinode close to the $\bar{M}$ point and greatest at the node, exhibiting the d-wave nature of the gap discussed in Sect. 2.2. This is seen more clearly in panels a - e, which show EDCs from the slices indicated by the yellow triangles in panel k. In panel a, the EDCs in the antinodal region of the BZ exhibit a gap of about 40 meV. The sharp peaks immediately below the gap are indicative of coherent excitations in the superconducting state. Moving along the k_x axis toward the Γ point, we see how these peaks broaden and the bonding band begins to open up. Moreover, the outermost peaks, belonging to EDCs in the Fermi surface, begin to disperse to smaller binding energies until they reach the Fermi energy (zero binding energy) in panel e.

Figure 5.3f–j shows the same sets of EDCs for the normal state data at 125 K. The EDCs in the antinodal region (panel f) are considerably broader than in the superconducting state, but still exhibit a gap of roughly the same magnitude as the superconducting gap. This is the pseudogap mentioned in Sect. 2.2. This pseudogap has a momentum dependence similar to the d-wave form of the superconducting gap seen below T_c, but this is not immediately apparent in these panels due to the wide energy scale, the broader peaks, and the proximity to the temperature $T^* \sim 130$ K, above which the pseudogap “fills in”.

Even though the data for this sample were collected up to a binding energy of 0.5 eV, the bonding band is seen to extend only up to about 0.4 eV in the spectra. Performing symmetry operations appropriate to a square BZ, these data can be extended to cover the entire BZ, as detailed below. The data are therefore complete in the sense that they cover almost the entire BZ in momentum space and the energy range of the occupied bonding band. Identical data sets were collected from this same sample at temperatures $T = 90, 110, 140$ K, as indicated by the blue squares in the phase diagram in Fig. 5.2b. In addition, similarly extensive data sets were considered for the other two samples, indicated by the red diamonds and black circles in Fig. 5.2b, but we do not show these for brevity.

We now discuss the method for extracting the spectral function from the data presented above. As discussed in Sect. 3.4, the ARPES signal can be expressed as [5]

$$I(\mathbf{k}, \omega) = I_0 f(\omega) A(\mathbf{k}, \omega) + B(\omega), \tag{5.5}$$

where $A(\mathbf{k}, \omega)$ is the imaginary part of the retarded, single-particle Green’s function—commonly known as the *spectral function*—$f(\omega)$ is the Fermi function, I_0 is a matrix element describing the variation of the intensity with incident photon energy and polarization, and $B(\omega)$ is a background signal. As shown schematically in Fig. 5.1, the *spectral function* is the foundation of nearly all the calculations presented below, so it had to be extracted from the ARPES signal.

The first step was to subtract the background spectrum. This signal was obtained differently for the thin films and the single crystal. The thin films naturally exhibit a weaker superlattice intensity and were measured with polarized synchrotron radiation, which can be tuned to suppress this signal even further [6]. For this reason, the background could be extracted directly from the data, by looking at regions of the BZ with no excitations. We chose the green patch in Fig. 5.2a and averaged all EDCs in this region to smooth out noise coming from low statistics. Since the background has been shown to be roughly independent of momentum [7], we subtracted the resulting spectrum from all EDCs in the data. For the single crystal, measured with unpolarized UV light, the superlattice signal is stronger, and there really is no region in momentum space where the signal is purely background (see Fig. 5.3k, l). For this reason, based on the general shape of the background inferred from the curves of Kaminski et al. [7] and the thin film data used in this work, as well as on the models for the background discussed by Kordyuk et al. [8], we adopted a simple model for the background consisting of a Fermi function with a linear tail at higher binding energies. Kordyuk et al. used a somewhat more complicated functional form, but the plots of their background spectra look quite similar to ours. More importantly, we have checked that the effects of varying this model background function on the spin susceptibilities and the Bethe-Salpeter eigenvalues are small.

The maximum binding energy for all thin film spectra was 0.64 eV, while for the single crystal data it was 0.51 eV. In the momentum window considered in this work, the bonding band does not disperse to such high binding energies, making them convenient reference points for normalization. Therefore, the background was first normalized at these energies to EDCs measured at all other momenta and then subtracted from them. This way, the background intensities and the matrix elements were removed simultaneously. We checked that the results presented throughout the remainder of this chapter were only slightly affected by variation of this high-energy cutoff. For example, for the OP80 thin film, we normalized the data at 0.5 eV instead of at 0.64 eV, with a resulting variation in the Bethe-Salpeter eigenvalue of only 3%. A more substantial variation was found when the binding energy was reduced to 0.3 eV. However, we believe normalizing at such low binding energies is incorrect, since the bonding band does disperse to such binding energies.

As noted, Bi2212 possesses a superlattice signal (diffraction replicas of the true bonding band, shifted from the latter by a vector $\pm(0.21, 0.21)\pi$) and an additional "shadow" band [9] (shifted from the main band by (π, π)) throughout the BZ (although this band was not discernible in any of the data used in this work). These features do not belong to the electronic structure of the CuO_2 planes and had to be removed. The background subtraction scheme outlined above partially lowered the spectral weight of these bands; the remaining signal was eliminated as follows: at every binding energy ω, we zeroed out the signal at momenta where it was 20% lower than the maximum intensity at that energy. In addition, we also zeroed out the spectra from momenta $k = k_F + \Delta k$ such that $\epsilon_k > 50$ meV, which is the extent to which symmetrization (see below) allows us to extend the spectra into the unoccupied states.

Having removed the background and the matrix element, we finally eliminated the Fermi function by means of a "symmetrization" procedure [10], in which the spectral function is assumed to be even with respect to the Fermi energy and the Fermi surface:

$$A(\mathbf{k_F} + \mathbf{k}_\perp, \omega) = A(\mathbf{k_F} - \mathbf{k}_\perp, -\omega), \tag{5.6}$$

where $\mathbf{k}_\perp$ is a momentum vector perpendicular to the Fermi surface at $\mathbf{k_F}$. In this approximation, which holds in the vicinity of the Fermi surface and for binding energies of the order of the superconducting gap, we may write the spectral function in terms of the ARPES intensity as

$$A(\mathbf{k}, \omega) = I(\mathbf{k}, \omega) + I(-\mathbf{k} + 2\mathbf{k_F}, -\omega), \tag{5.7}$$

where it is assumed the background has already been subtracted from the ARPES signal. This step is based upon an assumption of particle-hole symmetry, which allows us to extend the spectral function beyond the Fermi energy by about $\Delta = 50$ meV. In previous works, we have verified that p-h symmetry holds above [11] and below [12] T_c for under—and optimally doped Bi2212.

Given the finite electron analyzer acceptance angle and the kinematic restrictions inherent in photoemission spectroscopy, the momentum window measured in this work was the rectangular patch shown in Fig. 5.2a, consisting of 15 narrow rectangles representing discrete sample rotations (θ_i). To obtain the spectral function at all momenta, we first interpolated the data onto a rectangular grid, extracted the EDCs belonging to an irreducible octant of the zone (pink region in Fig. 5.2a), and reflected these about the symmetry axes of the BZ. For example, for the OP91 crystal, we selected the region in Fig. 5.3k and 5.3l bounded by the dashed lines, where the superlattice bands do not cross the main band. The resulting spectral function was normalized according to the sum rule [2]

$$\int_{-\infty}^{\infty} d\omega A(\mathbf{k}, \omega) = 1, \tag{5.8}$$

which ensured that the *total* spectral weight was the same for all temperatures and doping levels.

With these approximations, we extracted $A(\mathbf{k}, \omega)$ throughout the BZ (except for patches at the Γ point and near the zone corners) and for binding energies covering the bonding band. Figure 5.4 illustrates the procedure for the UD67 sample in the superconducting state. Panels a) and b) show the raw and processed Fermi surface scans, respectively. Note the superlattice signal in panel (a), which has been removed in panel (b). Panel (c) shows the raw EDCs along the path indicated by the colored dots. Panel d) shows the EDCs along the Fermi surface (from antinode to node) following subtraction of background and superlattice bands, and after symmetrization. The extracted spectral function covers energies in the range $[-0.64, 0.64]$ eV (where they go to zero), but we have displayed them in a reduced energy range to highlight the high energy resolution of the data, which allowed us to map the d-wave symmetry of the superconducting gap. Indeed, as we move along the Fermi surface starting from the antinode (red EDC in Fig. 5.4d), we see how the peak positions move toward the Fermi energy until they merge into a single peak at the node (green EDC). Similarly complete forms of the spectral function were obtained from ARPES data at all the points indicated in the phase diagram in Fig. 5.2b, but we have displayed the procedure for only one data set for illustration.

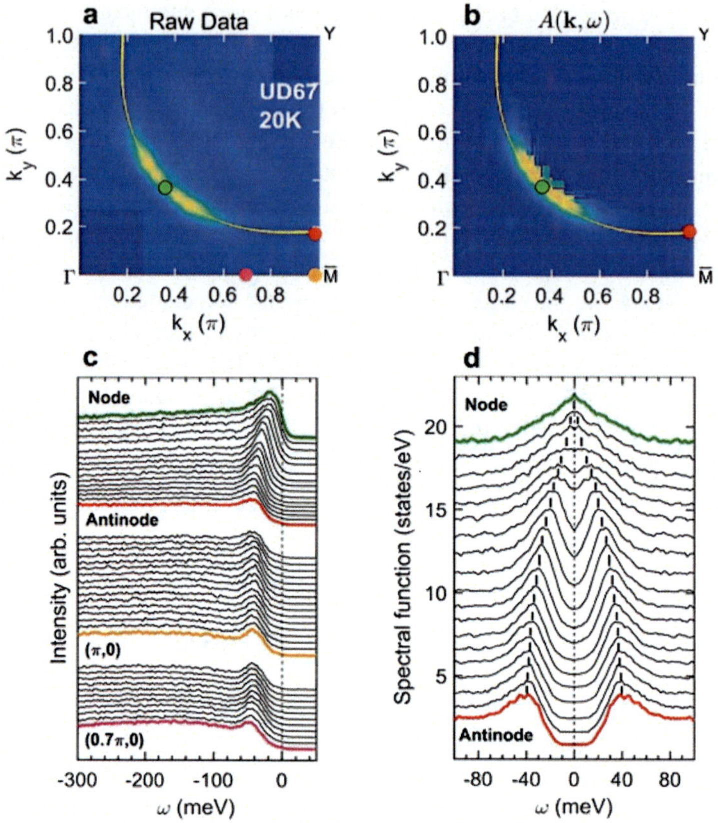

Fig. 5.4 Obtaining the spectral function from ARPES data from the UD67 sample. (**a**) Raw Fermi surface (after interpolation and reflection about symmetry axes). (**b**) Fermi surface after background subtraction, normalization, and symmetrization—the spectral function. (**c**) and (**d**) EDCs corresponding to panels (**a**) and (**b**), respectively. The colored EDCs correspond to the colored dots in (**a**) and (**b**) (adapted from Restrepo et al. Coupled electronic and magnetic excitations in the cuprates and their role in the superconducting transition. Commun. Phys. **5**, 45 (2022))

5.2 The Bare and Interacting Spin Susceptibilities in Real Frequencies

As noted above, we solved the pairing problem in Matsubara frequencies for numerical efficiency. Since the Matsubara response functions $\chi(\mathbf{q}, \omega_m)$ and $G(\mathbf{k}, \omega_n)$ entering the BSE are directly related to their real-frequency counterparts, our calculation began in real frequencies. The advantage of using ARPES data is that the spectral function $A(\mathbf{k}, \omega)$ is the building block for several other two-particle response functions observable in experiments. In particular, we can first calculate the spin susceptibility of a system with no electron-electron interactions—the *bare*

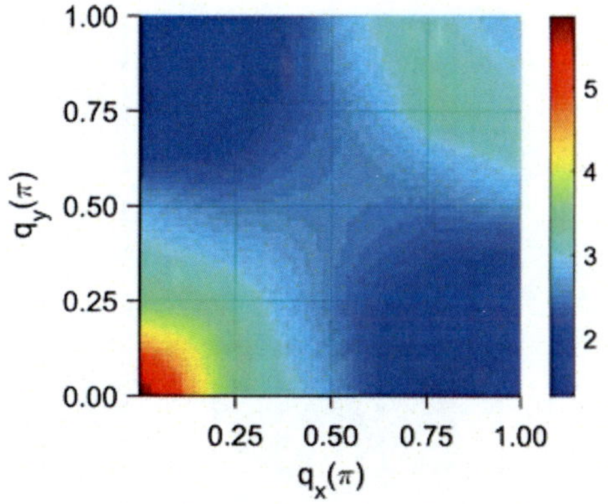

Fig. 5.5 Real part of the bare susceptibility at $\Omega = 0$ for an optimally doped Bi2212 sample at 140K (normal state). The color bar is in units of states/eV/f.u. (Reprinted from Restrepo et al. Coupled electronic and magnetic excitations in the cuprates and their role in the superconducting transition. Commun. Phys. **5**, 45 (2022))

susceptibility or *particle-hole bubble*—directly from the spectral function. Even though this is an extremely crude approximation to the spin susceptibility in the cuprates, we found it instructive to examine this quantity because (1) our results could be checked against similar calculations, performed using ideal, delta-function peaks for A, and (2) this comparison allowed us to identify an important issue with the calculation, as explained below. The bare susceptibility is given by

$$\chi_0(\mathbf{q}, \Omega) = \sum_{\mathbf{k}} \int_{-\infty}^{\infty} d\omega d\nu A(\mathbf{k}, \omega) A(\mathbf{k} + \mathbf{q}, \nu) \frac{f(\nu) - f(\omega)}{\Omega + \nu - \omega + i\delta}, \tag{5.9}$$

where $f(\omega)$ is the Fermi function and δ is a positive infinitesimal number. The momentum variable appears only in the spectral functions and the sum can be seen to have the form of a convolution. Such sums can be efficiently handled thanks to the Fourier convolution theorem. The energy sums were performed directly. The spectral functions were sampled over a 128×128 grid and the convergence factor δ was set to 5.5 meV, which is the lowest value for which smooth curves for χ_0 could be obtained. Figure 5.5 shows $\chi_0(\mathbf{q}, \Omega = 0)$ for an optimally doped Bi2212 sample in the normal state.[3] This quantity corresponds to the spin response of a system with no electron-electron interactions, and is therefore not comparable to the true spin response measured in the strongly correlated cuprates. Hence, the best comparison we can seek at this stage is with calculations based on non-interacting spectral functions for which the dispersion ϵ_k was derived from ARPES data from Bi2212, as the ones carried out in Ref. [4]. Similarly to that work, in Fig. 5.5 we see a pronounced, square-shaped signal enclosing the $\mathbf{q} = 0$ point, as well as an incommensurate response around $\mathbf{Q} = (\pi, \pi)$. However, there is also significant response at $\mathbf{q} = 0$, which is not seen in model calculations (Figure 1 of Ref. [4]).

[3] These data were taken by former group member Adam Kaminski and used in earlier publications, e.g. Refs. [13, 14].

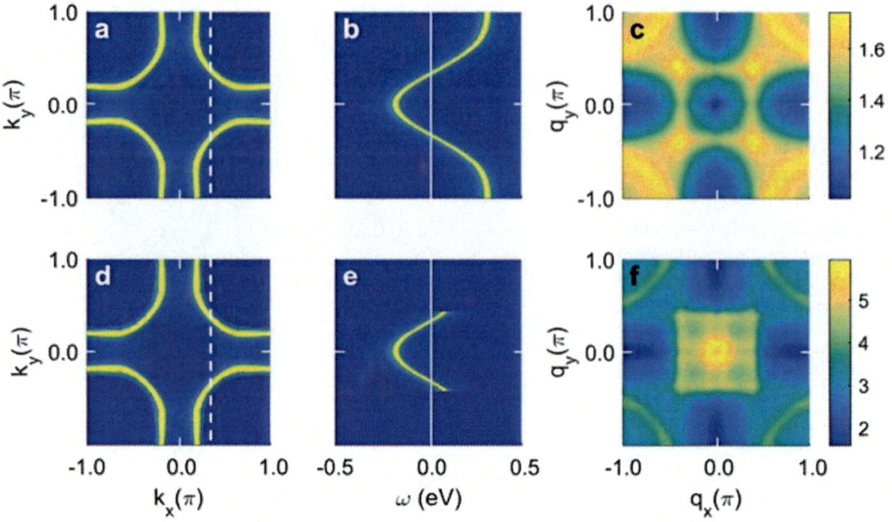

Fig. 5.6 Fermi surface, dispersion and bare susceptibility calculated from a Lorentzian spectral function with and without symmetrization. (**a**) Fermi surface of the true spectral function. (**b**) Dispersion along direction indicated by white vertical line in (**a**). (**c**) Calculated $\chi_0(\mathbf{q}, \Omega = 0)$. (**d**)–(**f**) Same as (**a**)–(**c**) but the spectral function has been symmetrized "by hand" and some patches of the BZ have been removed. Color bars in (**c**) and (**f**) are in units of states/eV/f.u (reprinted from Restrepo et al. Coupled electronic and magnetic excitations in the cuprates and their role in the superconducting transition. Commun. Phys. **5**, 45 (2022))

This response was found to be an artifact of the symmetrization procedure used to extrapolate the spectral function to positive binding energies.

To demonstrate the fictitious origin of the $\mathbf{q} = 0$ peak, we performed a calculation using model spectral functions with tight-binding fits for the dispersion of Bi2212.[4] We considered a normal state spectral function with a lorentzian lineshape of 20 meV width, sampled over a momentum grid of 128 × 128 points and energies ranging from −1.5 to 1.5 eV. The convergence factor δ was set to 10 meV. Figure 5.6a, b show the Fermi surface and the dispersion for the true spectral function (without symmetrization). The bonding band disperses (along the dashed line in panel a) all the way to the BZ boundary. The bare susceptibility (at $\Omega = 0$) for this spectral function (Fig. 5.6c) closely resembles similar calculations presented in Ref. [4]. Note the absence of spectral weight at $\mathbf{q} = (0, 0)$. The small, square-shaped signal around $\mathbf{q} = (0, 0)$ is a result of the scattering of electrons from the nearly straight antinodal regions of the Fermi surface. Similarly, the diagonal streaks close to the $(\pm\pi, \pm\pi)$ corners come from scattering between electrons from the nearly flat nodal regions. The second row shows similar calculations, but now

[4] The parametrization "tb2" in [4] was chosen to enhance the incommensurate response around $\mathbf{Q} = (\pi, \pi)$. It also yielded the χ_0 which most closely resembles the one obtained from real data.

the spectral function has been multiplied by the Fermi function and symmetrized, to simulate the analysis we did with experimental data. The dispersion in panel e does not cover the entire BZ because particle-hole symmetry is only assumed to hold for small energies (in our case, 50 meV above the Fermi energy). We see that the resulting bare susceptibility (panel f) has significant spectral weight at the zone center (note also the difference in color bar scales). The artifact is thus a direct consequence of not having access to the complete spectral function (including the unoccupied states), a shortcoming which can only partially be mitigated through symmetrization.

Now, the quantity which is actually detected in experiments on the cuprates, and which ultimately enters the BSE, is the *interacting* spin susceptibility, which takes electron correlations into account through the coupling function U_q (see below). Unfortunately, the artifact mentioned above also plagues this quantity, approximated via RPA as [15]:

$$\chi(\mathbf{q}, \Omega) = \frac{\chi_0(\mathbf{q}, \Omega)}{1 - U_q \chi_0(\mathbf{q}, \Omega)}. \tag{5.10}$$

Indeed, if U_q is momentum-independent, χ possesses an anomalously strong peak near $\mathbf{q} = 0$, which is not seen in INS measurements. In Fig. 5.7 we compare the magnetic dispersion for χ calculated with a momentum-independent coupling U_0 from $\mathbf{q} = (\pi/2, \pi/2)$ to $\mathbf{q} = (3\pi/2, 3\pi/2)$ (as is frequently reported in the literature) with the same quantity plotted from $\mathbf{q} = (0, 0)$ to $\mathbf{q} = (2\pi, 2\pi)$. The data correspond to an optimally doped Bi2212 sample in the superconducting

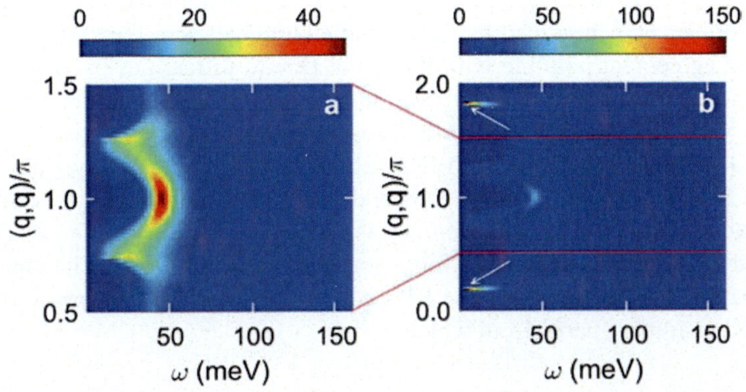

Fig. 5.7 Symmetrization artifacts in the interacting spin susceptibility of near-optimally doped Bi2212. (**a**) ARPES-derived magnetic excitation spectrum along diagonal direction from $(0.5\pi, 0.5\pi)$ to $(1.5\pi, 1.5\pi)$ (**b**). Same as (**a**), but in a momentum range $(0, 0) - (2\pi, 2\pi)$. Note the difference in colorbar scales (states/eV/f.u.) (reprinted from Restrepo et al. Coupled electronic and magnetic excitations in the cuprates and their role in the superconducting transition. Commun. Phys. **5**, 45 (2022))

state.[5] In the restricted momentum view (panel a), U_0 yields a spin response with rich momentum structure, resembling the famous "hourglass" shape reported in the literature on other cuprates like YBCO. However, the wider momentum view (panel b) reveals two peaks near $\mathbf{q} = (0, 0)$ and $\mathbf{q} = (2\pi, 2\pi)$ a factor of 5 larger than the rest of the signal. This low-$\mathbf{q}$ signal, which is not detected in INS experiments, is the manifestation of symmetrization on the interacting spin susceptibility.

To circumvent this problem, we first note that the effective AFM coupling U_q is a high energy scale ($\sim$ 1 eV) not provided either by the RPA nor by any low-energy spin-fluctuation-based theory (where high energy scales are projected out of the Hamiltonian) [16]. Since INS experiments measure Imχ (modulo μ_B^2, a matrix element, and a Bose occupation factor) [15], we can choose U_q so that the calculated dynamic susceptibility approximately matches the measured one. Moreover, the choice can be such as to minimize the fictitious signal near $\mathbf{q} = 0$ (Fig. 5.7b). In particular, we note that the function

$$U_q = -\frac{U_0}{2}(\cos q_x + \cos q_y) \tag{5.11}$$

suppresses the spin response near $\mathbf{q} = 0$. This form of U_q is commonly employed in the literature and arises naturally in the *t*-*J* model, where it represents a super-exchange interaction between magnetic moments centered at Cu sites in the CuO_2 planes [3, 4, 17, 18]. We emphasize that such a choice of U_q was not made to favor spin-fluctuation-induced superconductivity models, but was rather motivated by the fact that it rectifies the small-$\mathbf{q}$ intensity problem and yields a spin susceptibility in better agreement with experimental observations. We also note that U_0 is the same spin-fermion coupling as that in Eq. 5.3.

Having identified and rectified these issues, the calculation of the interacting spin susceptibility is then guided by the simple requirement that the choice of U_0 gives the best agreement possible with INS data. Due to difficulties in fabricating large enough Bi2212 samples and the low counting rates of INS measurements, neutron data from Bi2212 are scarce. For this reason, we will mostly compare our ARPES-derived spin susceptibilities with INS spectra from the extensively studied cuprate $YBa_2Cu_3O_{6+x}$ (YBCO). In the superconducting state, the most salient feature in the spin excitation spectrum of the cuprates is the so-called "spin resonance" (Sect. 2.4), located at the commensurate wave-vector $\mathbf{Q} = (\pi, \pi)$ [19, 20]. The emergence of this feature can be seen from the RPA expression (Eq. 5.10), by resolving χ_0 into real and imaginary parts, from which we get

$$\chi''(\mathbf{q}, \Omega) = \frac{\chi_0''(\mathbf{q}, \Omega)}{(1 - U_q \chi_0'(\mathbf{q}, \Omega))^2 + (U_q \chi_0''(\mathbf{q}, \Omega))^2}. \tag{5.12}$$

[5] These data were collected from the same sample as those in Fig. 5.5 and are used here only for illustration purposes.

To get a resonance at $\mathbf{Q} = (\pi, \pi)$ and an energy Ω_0, we first note that $U_q = U_0$ at this momentum vector. The value of U_0 that maximizes $\chi''(\mathbf{Q}, \Omega_0)$ is given by

$$U_0 = \frac{\chi_0'(\mathbf{Q}, \Omega_0)}{(\chi_0'(\mathbf{Q}, \Omega_0))^2 + (\chi_0''(\mathbf{Q}, \Omega_0))^2}. \tag{5.13}$$

Then, the only input parameter is the position of the resonance peak, Ω_0. The energy of this resonance throughout the phase diagram of YBCO has been extensively reported; however, due to technical limitations, such information is lacking for underdoped Bi2212. For this reason, we used the empirical relation [21, 22]

$$\Omega_0 = 5.4 k_B T_c \tag{5.14}$$

(where T_c is the superconducting transition temperature) to fix the resonance energies for the three samples ($\Omega_0^{67} = 32$ meV, $\Omega_0^{80} = 37$ meV, and $\Omega_0^{91} = 41$ meV), allowing us to extract the RPA parameter U_0 and calculate the interacting spin susceptibility.

Equation 5.14 suggests that the resonance energy varies with doping but not with temperature, and then only in the superconducting state, where this feature is observed. At low doping, INS has also shown that, for $T > T_c$, the resonance simply broadens and loses intensity, but its energy remains relatively fixed with temperature, especially in the temperature range considered here [20, 23–25]. This is correlated with the fact that the antinodal spectral gap seen in ARPES (which persists above T_c in the pseudogap regime) is nearly constant up to the pseudogap scale T^* (Fig. 5.2b), where it fills in.

Whereas the above viewpoint is fully consistent with ARPES and INS data in *underdoped* YBCO, optimal doping presents a more serious challenge in that, to our knowledge, there is only one study where INS detected magnetic signal above T_c at this doping level [26]. (Interestingly, the data reported in this work belonged to the cuprate Bi2212, thanks to recent advances in the fabrication of sufficiently large, high-quality crystals.) We nevertheless follow the same procedure, motivated by the following three observations: (1) despite the error bars, Ref. [26] found that the overall shape of the magnetic dispersion (intensity versus energy and momentum) is roughly the same in the pseudogap and SC states of Bi2212. In particular, they found a commensurate feature above T_c at the same energy as the resonance below T_c. (2) While their observation in YBCO has eluded INS presumably due to high phonon backgrounds and low counting statistics, NMR shows unequivocally that there are indeed magnetic excitations in optimally doped YBCO above T_c [27, 28]. Moreover, Fong et al. [29] have estimated an upper bound for this signal for $YBa_2Cu_3O_7$ ($\sim$ 30 states/eV/formula unit) which is consistent with the values obtained in our calculations below. (3) In the itinerant electron picture adopted here (Eq. 5.10), the energy of the commensurate feature is correlated with the ARPES spectral gap, which is still sizable at optimal doping and does not change in magnitude when crossing T_c [30]. From these considerations, we can say that Eq. 5.14 gives the resonance energy at doping levels indicated by T_c, but that this feature does not

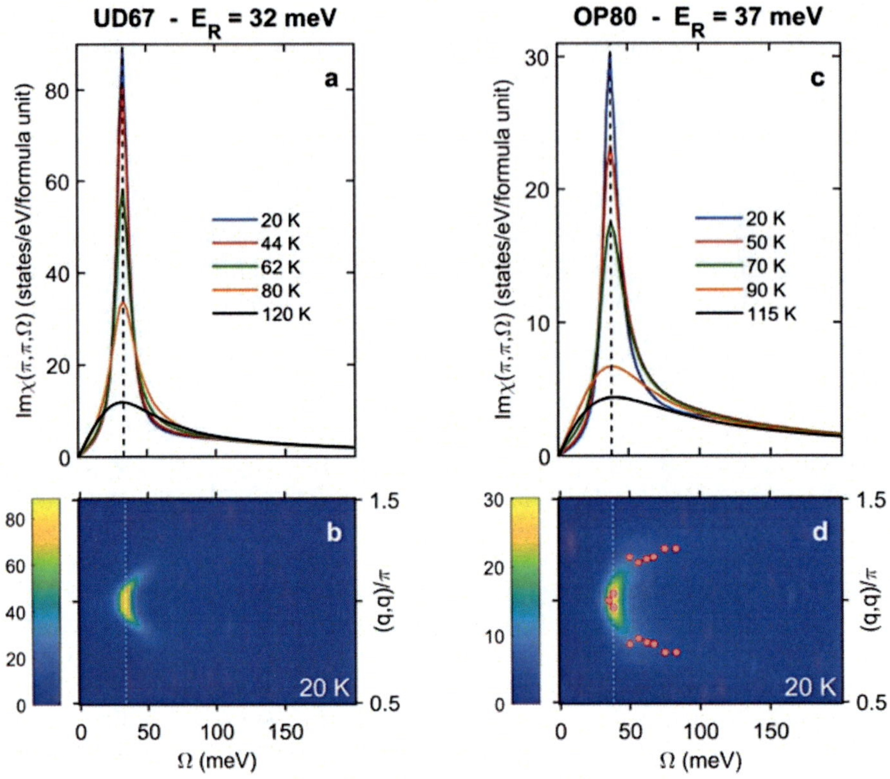

Fig. 5.8 (**a**), (**c**) Calculated spin susceptibility for both thin films (UD67 and OP80) as a function of energy at the commensurate wave vector (π, π) for various temperatures. (**b**), (**d**) Dispersion of magnetic excitations along the diagonal direction at 20 K for UD67 and OP80 samples, respectively. Pink symbols in (**d**) represent INS measurements only above the resonance energy from optimally doped Bi2212 at 10 K (adapted from Ref. [26]). Color bars in (**b**) and (**d**) are in units of (states/eV/formula unit) and dashed lines indicate the resonance energies (reprinted from Restrepo et al. Coupled electronic and magnetic excitations in the cuprates and their role in the superconducting transition. Commun. Phys. **5**, 45 (2022))

disperse with temperature (at least below the maximum temperature considered in this work, $T = 140$ K). In Sect. 5.4 we discuss the consequences of relaxing this assumption on the Bethe-Salpeter eigenvalues.

In this way, we calculated the dynamic spin response while extracting the spin-fermion coupling for each ARPES data set. Figure 5.8a,c shows our results for Imχ at the commensurate wave-vector as a function of energy. For brevity, only the results for the two thin films are shown, but similar results were obtained for the OP91 crystal (see Appendix C). By construction, there is a sharp resonance at the energies given by Eq. 5.14 well in the superconducting state (20 K). The commensurate feature decreases in intensity and exhibits considerable broadening with increasing temperature. Moreover, comparing panels a and c, we also see that the peak intensities are higher at lower doping, which has also been seen in INS experiments on YBCO [25]. This trend is more evident in Fig. 5.10a below, which shows the peak intensity (this time for the three samples) as a function of temperature (the lines are guides to the eye; in reality, the curves should flatten out at the higher pseudogap temperature scale T^* [23] shown in Fig. 5.2b). Also, the resonance peaks' widths deep in the superconducting state are $\sim$ 15 meV, in agreement with INS data from optimally doped Bi2212 [19].

Even though the emergence of the resonance peak from Eq. 5.12 is seen here as a simple mathematical consequence of the RPA (Eq. 5.10), it is important to remember that in this approximation the spin susceptibility is expressed as an infinite summation of terms with increasing powers of the particle-hole bubble. As such, there is an implicit assumption that this mode is an $S = 1$ particle-hole bound state, also known as a *spin exciton*. Our results in Fig. 5.8a, c suggest that this magnetic excitation is sharp below T_c because electronic excitations are gapped in the superconducting state and, thus, there are less states for the resonance to decay into. Above T_c, the gap "softens" in the pseudogap regime and there are more low-energy states which effectively dampen the resonance. Thus, the broadening of the resonance with increasing temperature can be viewed as a consequence of the increased scattering of electrons off spin fluctuations. While the spin-exciton picture adopted here is consistent with INS data, there are other proposals for the physical origin of the spin resonance, some of which have led to findings similar to ours (see Ref. [31] and references therein). For this reason, we do not insist on the relative merits of the spin-exciton interpretation. Instead, we stress that the damping mechanism described above is common across different proposals for the origin of the spin resonance [16, 31, 32] and, therefore, the spin-fermion interaction can reasonably account for the temperature dependence of the cuprate spin response measured in INS. We also emphasize that, operationally, the spin-exciton picture yields spin susceptibilities in reasonable agreement with INS data while allowing us to extract the spin-fermion coupling (see below). As far as the solution of the BSE is concerned, this is all that is required.

We now turn to the momentum dependence of the magnetic excitations. We note that our calculations reproduce the ubiquitous "upward branch" ($\Omega > \Omega_0$) of the "hourglass" (Sect. 2.4), observed in Bi2212 and all other cuprate families (Fig. 5.8 b, d). In particular, note the approximate coincidence of our calculated $\mathrm{Im}\chi(\mathbf{q}, \Omega)$ with the experimentally determined dispersion of magnetic excitations in optimally doped Bi2212 at 10 K (pink symbols in Fig. 5.8d). We found a similar agreement with INS data in the analysis of the OP91 single crystal in the normal state, as shown in Appendix C. Aside from the upward dispersing branch, it is interesting to note the two weak, almost linearly dispersing features on either side of the resonance in Fig. 5.8d. To better appreciate these features, we turn to Fig. 5.9, which is the same as Fig. 5.8d but in the reduced energy range [0, 100] meV. The white curve is the locus of minima of the quantity $E_{\mathbf{k}} + E_{\mathbf{k+q}}$, where $\mathbf{q}$ runs along $1/2(\pi, \pi) \rightarrow 3/2(\pi, \pi)$, $E_{\mathbf{k}}$ is the gapped dispersion $E_{\mathbf{k}} = \sqrt{\epsilon_{\mathbf{k}}^2 + \Delta_{\mathbf{k}}^2}$, and $\Delta_{\mathbf{k}}$ is the d-wave gap. We have taken $\Delta_0 = 35$ meV as the antinodal gap and a parametrization of

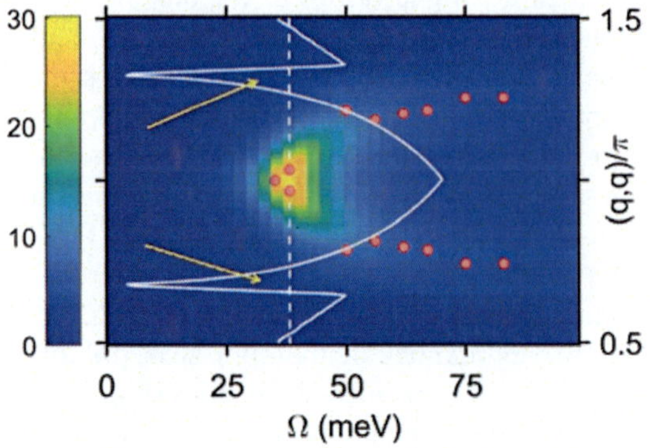

Fig. 5.9 Dispersion of magnetic excitations in the OP80 sample at 20 K. The white curve represents the lower threshold for particle-hole excitations. The yellow arrows indicate the narrow regions of allowed non-resonant excitations (reprinted from Restrepo et al. Coupled electronic and magnetic excitations in the cuprates and their role in the superconducting transition. Commun. Phys. **5**, 45 (2022))

the dispersion which gives the Fermi surface in closest agreement with our data for the OP80 sample. Non-resonant magnetic excitations are allowed only to the right of this curve. Any excitations occurring in the gapped region (like the spin resonance) are protected against decaying into particle-hole pairs by the gap; hence, they are sharp in momentum and energy. The complicated shape of this "particle-hole threshold" is controlled by the d-wave gap and the shape of the quasiparticle Fermi surface. Note the two regions of allowed excitations indicated by the arrows. These regions are located near vectors $\mathbf{q_1} = 0.68(\pi, \pi)$ and $\mathbf{q_2} = 1.33(\pi, \pi)$, which correspond to the two possible node-node scattering vectors in the Fermi surface. Thus, these features can be understood as signatures of Fermi surface nesting in the particle-hole bubble χ_0.

The "lower branch" ($\Omega < \Omega_0$) of the magnetic excitations, which does not appear to be common across the cuprate family [33, 34], is not reproduced in our calculations. It is possible that other choices of the function U_q may capture this low-energy structure, but our focus was rather on reproducing universal features and, more importantly, tracking the temperature and doping dependence of the (π, π) peak intensity and energy. Moreover, it has been shown [3, 4] that the spin susceptibilities obtained from RPA-based calculations tend to be strongly sensitive to the form of the underlying dispersion ϵ_k, so that, for example, one can reproduce the lower branch with one set of tight-binding parameters, but not with a different set. When using real data, there is an analogous instability in the procedure related to how the ARPES data are extended to cover the entire BZ, which can ultimately alter the resulting dispersion. Finally, this low-energy structure takes up only a small fraction of the magnetic spectral weight and, hence, is not expected to significantly affect the solution of the BSE. For these reasons, we have not insisted on reproducing this finer structure. The upper branch, which is a robust feature of our calculations, is more important, as it takes up a substantial portion of the spectral weight, as we shall discuss next.

Since our main goal was calculating the leading eigenvalue of the BSE, which is a dimensionless number, it is imperative that the factors entering this equation are expressed in absolute units; that is, the Green's functions are to be expressed in (states/eV) and the spin response in (states/eV/formula unit). The correct normalization of the Green's function follows automatically from the normalization of the spectral function according to Eq. 5.8. For the spin response, there is an analogous sum rule for the total magnetic spectral weight, according to which the energy-momentum integral of Imχ is independent of temperature [35]. The actual value for this integral can only be determined for a system of fixed magnetic moments; i.e. a statically ordered magnet. Since this is not an appropriate description for the doped cuprates, the best we can do is to compare our results with spectral weights obtained from INS data from cuprates at similar doping levels as the ones considered in this work. To compare with the INS intensities reported in the literature, we first multiply our calculated spin susceptibilities by the matrix element $2\mu_B^2$ [4]. The energy-momentum integral of Im$\chi(\mathbf{q}, \Omega)$ gives a total fluctuating moment of $\langle m^2 \rangle \sim 0.43\mu_B^2$ for the UD67 sample and $\langle m^2 \rangle \sim 0.39\mu_B^2$ for OP80, both varying weakly with temperature, even above and below T_c. Moreover, for the UD67 sample at 20 K, the total spectral weight taken up by the resonance (integrated over all momenta and from 0 to 70 meV) is about $0.16\mu_B^2$, whereas the remaining spectral weight (up to 200 meV), $\sim 0.27\mu_B^2$, is taken up by high-energy magnetic excitations. These values compare reasonably well with the corresponding figures for $YBa_2Cu_3O_{6.6}$ (with a similar T_c) [36] and suggest that the high-energy structure of the spin response must be included in a description of the magnetically mediated pairing interaction.

The last aspect of these calculations we wish to discuss is the temperature and doping dependence of the spin-fermion energy U_0, summarized in Fig. 5.10b. We see that there is a noticeable enhancement of this energy approaching T_c from above, for both doping levels. In particular, note the enhancement in the UD67 sample at $T = 62$ K (slightly below T_c), where $U_0 = 737$ meV. This is consistent with short-range AFM correlations becoming stronger as the temperature is lowered toward T_c [37]. It is also interesting to note that the coupling constants are larger in the underdoped sample, presumably because of stronger spin fluctuations near the AFM phase. Below T_c, U_0 either decreases or plateaus with decreasing temperature. NMR has detected similar tendencies in the ratio of planar copper to oxygen spin-lattice relaxation rates ($^{17}T_1/^{63}T_1$) below T_c of $YBa_2Cu_3O_7$ [27, 28], which is also taken as a measure of AFM correlations. However, uncertainties coming from the anomalous Green's function (see Appendix D) precluded further attempts to establish the precise temperature dependence of AFM correlations below T_c from our analysis.[6]

[6] In Fig. 5.10b, due to their proximity to T_c, we have treated the data from the UD67 sample at 62K, from the OP80 sample at 70 K, and from the OP91 sample at 90K, as representative of the normal state. Therefore, the U_0 values at these three points do not include the effect of the anomalous Green's function.

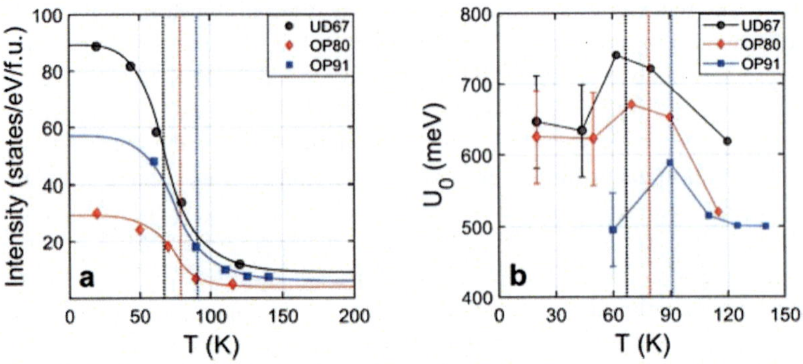

Fig. 5.10 (**a**) Temperature dependence of commensurate peak intensity for the three Bi2212 samples. (**b**) Temperature dependence of the spin-fermion coupling U_0 for all samples. The vertical dotted lines indicate T_c for the UD67 (black), OP80 (red), and OP91 (blue) samples. The large variations in U_0 below T_c represent our estimates of the uncertainties coming from the anomalous Green's function (see Appendix D)

5.3 Solving the BSE in Matsubara Frequencies

Finding ourselves in possession of the response functions $A(\mathbf{k}, \omega)$ and $\text{Im}\chi(\mathbf{q}, \Omega)$ and the spin-fermion coupling U_0, we are finally in position to solve the pairing problem defined by Eqs. 5.1–5.3. In terms of the workflow diagram in Fig. 5.1, in the last sections we have described the steps colored in red and blue. This section covers the final (pink) steps in solving the BSE, which we repeat here for convenience:

$$\lambda(T)\Phi(\mathbf{k}, \omega_n, T) = -\frac{3U^2}{2}\frac{T}{N}\sum_{\mathbf{k}',\Omega_n} \chi(\mathbf{k}-\mathbf{k}', i\omega_n - i\Omega_n, T) \times G(\mathbf{k}', i\Omega_n, T)G(-\mathbf{k}', -i\Omega_n, T)\Phi(\mathbf{k}', \Omega_n, T). \quad (5.15)$$

Since the BSE is a linearized gap equation, it is only valid above T_c; therefore, the rest of the discussion will be limited to the normal state. The Matsubara forms for the Green's function and the spin susceptibility follow directly from the spectral function and $\text{Im}\chi(\mathbf{q}, \Omega)$ by virtue of the Lehmann representations [2]

$$G(\mathbf{k}, i\omega_n, T) = \int d\omega \frac{A(\mathbf{k}, \omega, T)}{i\omega_n(T) - \omega} \quad \text{and} \quad \chi(\mathbf{q}, i\omega_m, T) = -\frac{1}{\pi}\int d\Omega \frac{\text{Im}\chi(\mathbf{q}, \Omega, T)}{i\omega_m(T) - \Omega}, \quad (5.16)$$

where explicit temperature dependencies have been included to emphasize the fact that we used spectral functions and spin susceptibilities derived from data at various temperatures. The number of fermionic Matsubara frequencies was chosen so that $G(\mathbf{k}, i\omega_n)$ drops below 5% of its maximum value at the maximum frequency, which was fixed at $\pm\omega_{n,max} = \pm 15$ eV for all calculations, so as to not introduce fictitious temperature dependences. $\chi(\mathbf{q}, i\omega_m)$ decays as ω_m^{-2} and so could be sampled only up to ± 3.5 eV. The energy ranges for both G and χ were varied by up to 20%, with no appreciable effect on the solutions to the BSE.

For $T > T_c$, λ is typically less than unity; therefore, we focus only on the largest (or leading) eigenvalue of $\hat{O}$. To this end, we used the power method [38], which is an iterative procedure to calculate the maximum eigenvalue of an operator (see Appendix A for details). We first present our results for the obtained eigenfunctions for the thin films in the critical region near T_c (additional results for the OP91 crystal can be found in Appendix C). Figure 5.11a, b shows the calculated eigenfunction Φ at $\omega_{n=0}$ as a function of momentum along the path indicated in the inset of Fig. 5.11b. As expected, this function, which reflects the symmetry of the superconducting gap, changes sign when crossing the BZ diagonals. Note the close similarity between these curves and the pure d-wave function $\frac{1}{2}[\cos(k_x) - \cos(k_y)]$ (black dashed line in panels a and b). The calculated Φ also allows us to recover the superconducting gaps measured in ARPES along the Fermi surface (panel e), as

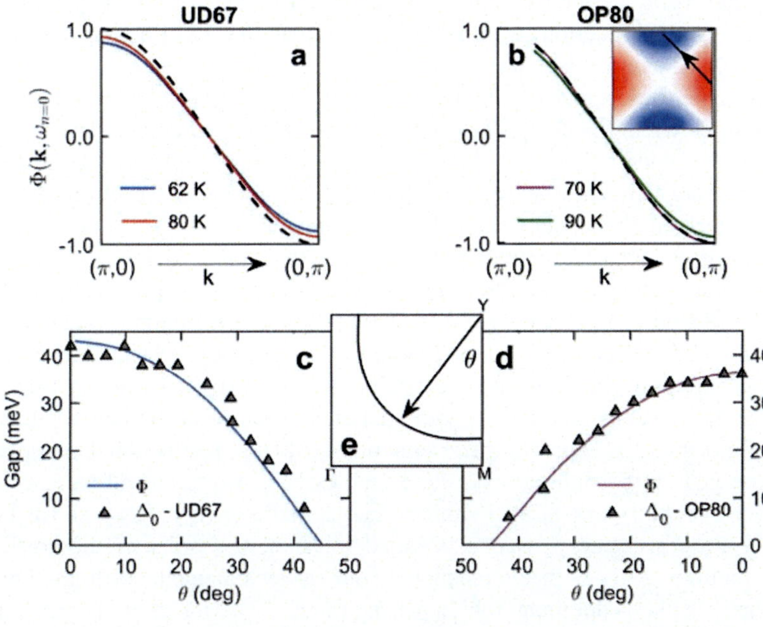

Fig. 5.11 (**a**) Momentum dependence of eigenfunction Φ along the $(\pi, 0) \rightarrow (0, \pi)$ direction (as shown in the inset of (**b**)) for the UD67 sample at $T = 62$, 80 K. (**b**) Same as (**a**) but for the OP80 sample at $T = 70$, 90 K. (**c**) Comparison of ARPES gap (Δ_0, gray triangles) for UD67 at 20 K with the corresponding $\Phi(\mathbf{k}, k_B T)$ (at 62 K, solid line) along the Fermi surface shown in (**e**). (**d**) same as (**c**), but the ARPES gap is for the OP80 sample at 20 K and the pairing eigenfunction was calculated from the OP80 data at 70 K. The black, dashed line in (**a**) and (**b**) is the d-wave function $\frac{1}{2}[\cos(k_x) - \cos(k_y)]$ along the path $k_y = \pi - k_x$ (adapted from Restrepo et al. Coupled electronic and magnetic excitations in the cuprates and their role in the superconducting transition. Commun. Phys. **5**, 45 (2022))

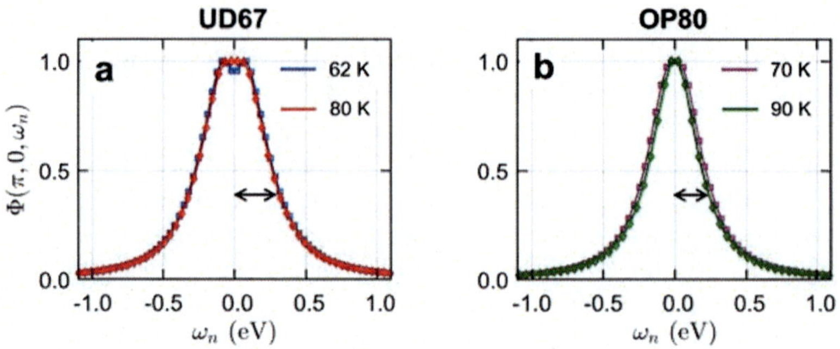

Fig. 5.12 Energy dependence of the order parameter Φ at $(\pi, 0)$ for (**a**) the UD67 sample at $T = 62, 80$ K and (**b**) the OP80 sample at $T = 70, 90$ K. (Adapted from Restrepo et al. Coupled electronic and magnetic excitations in the cuprates and their role in the superconducting transition. Commun. Phys. **5**, 45 (2022))

shown in panels c (for UD67 at 62 K) and d (for OP80 at 70 K), where the ARPES gap values (gray triangles) were obtained for both films at 20 K. In panels c and d, the pairing eigenfunctions were re-scaled to match the ARPES gaps at $\theta = 0°$ and the gap values were adapted from [30].

The energy dependence of these eigenfunctions at the momentum coordinate $(\pi, 0)$ is shown in Fig. 5.12. The curves are even in frequency, and have a characteristic decay scale of about 250 meV. Since physical observables cannot be measured in Matsubara frequency space, we compare our results with Quantum Monte Carlo simulations on the widely studied, two-dimensional Hubbard model, which are performed in Matsubara frequencies. While the Hubbard model is certainly a minimal description of the electronic structure of the CuO_2 planes in the cuprates, numerical studies have shown that it captures some of the cuprate phenomenology, like an AFM order in the undoped system [39], pseudogap behavior [40], d-wave superconductivity [41–43] and even stripe phases [44]. Of particular interest here are the calculations of the energy-momentum structure of the pairing interaction. In solutions of the Bethe-Salpeter equation (with the interaction V_{SF} replaced by a more general pairing vertex), these studies have found eigenfunctions Φ with d-wave symmetry and with a similar characteristic energy decay as we found in this work. This energy scale was found to be of the order of the particle-hole continuum, $\sim 2J$ [42], in agreement with our earlier statement that high-energy spin fluctuations play a dominant role in pairing (Sect. 5.2). Moreover, if we interpret the decay scale obtained from our calculations in this way, we obtain values of the exchange energy in the range 120 meV$< J <$ 130 meV, in agreement with values obtained from INS for the cuprates $La_{2-x}Sr_xCuO_4$ and $YBa_2Cu_3O_{6+x}$ [35]. We may also tentatively add that the effective exchange energy is larger in UD67 than in OP80 (see arrows in Fig. 5.12a, b), reflecting the fact that exchange correlations increase with proximity to the AFM phase.

The energy decay of the order parameter Φ implies a retardation in time of the pairing interaction of order $(2J)^{-1}$, consistent with recent ultrafast optical

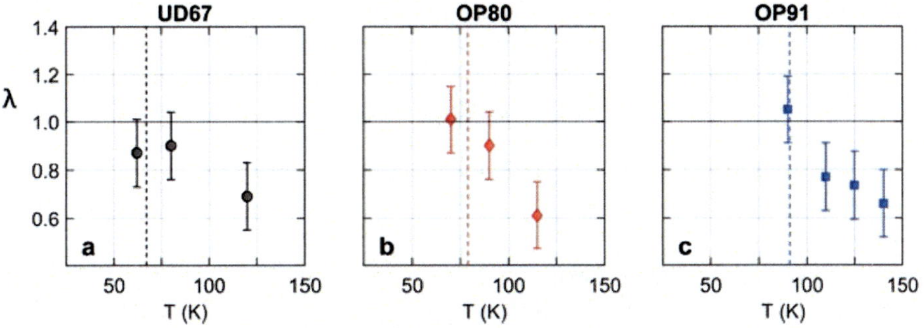

Fig. 5.13 Temperature dependence of leading Bethe-Salpeter eigenvalues for three Bi2212 samples. The dotted lines indicate the *true* critical temperatures for the three samples. The error bars are dominated by the uncertainty in the coupling energy U_0, coming from the finite resolution of INS measurements and possible deviations of the (π, π) peak energy from the value given by Eq. 5.14 (see Sect. 5.4) (reprinted from Restrepo et al. Coupled electronic and magnetic excitations in the cuprates and their role in the superconducting transition. Commun. Phys. **5**, 45 (2022))

spectroscopy measurements, which reported retardation times of $\sim$ 16 fs in an Yttrium-doped Bi2212 sample [45]. This result suggests that the dynamics of the pairing interaction is set by the spectrum of magnetic excitations probed in INS, which for the cuprates extends over an energy $2J \sim 250$ meV. As such, the energy scale inferred from Φ is also consistent with the notion that Cooper pairing in the cuprates arises from the exchange of spin fluctuations. In contrast, it is interesting to note that in RVB-type theories (see Introduction), the energy scale of the pairing interaction is set by the larger Mott-Hubbard gap ($\sim$ 2 eV), so the pairing interaction is essentially instantaneous in these frameworks.

To summarize the findings displayed in Figs. 5.11 and 5.12, we stress that, while the above comparisons with Hubbard model calculations were motivated by the lack of experimental data in Matsubara frequencies, the eigenfunctions Φ calculated in this work were found to be in agreement with (1) the d-wave pairing gap obtained from ARPES, (2) the cuprate exchange energies inferred from INS, and (3) the time retardation of the pairing interaction observed in optical spectroscopy measurements.

We then proceeded to examine the eigenvalues corresponding to the eigenfunctions discussed above. The main result of this work is shown in Fig. 5.13, which summarizes the temperature dependence of these eigenvalues in the normal state (including three results below, but close to, T_c, as noted in Sect. 5.2). We first note the temperature dependence of the eigenvalue; for the three samples, it increases as T is lowered toward T_c. This temperature dependence is a direct consequence of using temperature-dependent, experimental spectral functions. Again, some of the Hubbard model calculations referenced above (Refs. [37, 41–43]) are consistent with these findings. As in these works, we attribute the temperature dependence of λ to an increase in short-range AFM correlations as the temperature is lowered toward T_c. In this regard, we note that Mishra et al. [14] argued that, since above T_c the low-energy electronic states are suppressed by the pseudogap, there are less electrons at the Fermi surface to bind into Cooper pairs. Consequently, they concluded that the spin-fermion interaction (or any other pairing "glue") could not mediate high-T_c superconductivity. On the other hand, Maier et al. [37] argued that, while there was indeed a depletion of electronic states at the Fermi surface, there was a concomitant increase in AFM correlations which could compensate for this effect and yield a high-temperature superconducting transition. Our results are in line with this latter proposal.

In panel (a), the eigenvalue is seen to decrease somewhat below T_c, in contrast to the monotonic behavior in panels (b) and (c). We speculate that this may be related to the larger superconducting gap in the UD67 sample compared to the OP samples (see Fig. 2.3). This larger gap means that the Cooper pairs are more strongly bound, so the contribution from the anomalous Green's function (see Appendix D), which was neglected in the present calculation, may need to be accounted for. Be that as it may, the corrections in λ coming from such a procedure are expected to be smaller than the error bars in Fig. 5.13, so pursuing this issue further seems rather pedantic.

Our main finding is that, to the accuracy of the present calculation (see next section), the eigenvalues are essentially equal to unity near the T_c of each sample. These results suggest that *spin fluctuations are sufficiently strong to mediate high-temperature, d-wave superconductivity in Bi2212 at optimal and sub-optimal doping.*

Going back to the introduction, it is instructive to ask why Mishra et al. obtained eigenvalues of about 0.2, independent of temperature, even though they followed a procedure similar to ours. We believe this could due to the following reasons:

- As noted in the introduction, they considered ARPES data from an optimally doped Bi2212 sample only at 140 K. Furthermore, they argued that the temperature dependence of the eigenvalues could be approximately accounted for by the Matsubara frequencies, which depend on temperature. In other words, they constructed both the Matsubara Green's function and spin susceptibility according to Eqs. 5.16 but using $A(\mathbf{k}, \omega)$ and $\text{Im}\chi(\mathbf{q}, \Omega)$ at a single temperature. This may explain why they obtain a T-independent eigenvalue.
- The low value $\lambda \sim 0.2$ may be due to the form of the spin susceptibility Mishra et al. obtained in real frequencies. Even though their $\text{Im}\chi$ had a maximum intensity of about 5 (states/eV/formula unit), which is similar to what we obtained for the similarly doped thin film at 115 K (see black curve in Fig. 5.8c), the spectral weight was mostly concentrated at incommensurate wavevectors away from (π, π). In contrast, the spin susceptibility obtained here is peaked at the commensurate wavevector. If we recall our discussion in Sect. 2.4, a d-wave superconducting solution to the gap equation is favored if the pairing interaction (which is proportional to $\text{Im}\chi$) is peaked at (π, π). We should emphasize here that our form of $\text{Im}\chi$ not only enhances d-wave superconductivity, but is also in better agreement with the spin susceptibility measured in recent INS experiments on optimally doped Bi2212 above T_c [26].

5.4 Uncertainties in the Eigenvalues

The leading eigenvalue in the BSE depends on the spin-fermion coupling through the factor U_0^2 and the denominator of the RPA expression for the spin susceptibility. This coupling, in turn, was fixed from the empirical relation Eq. 5.14. Therefore, we estimated the uncertainty in the eigenvalue by relaxing the above relation and introducing the uncertainty $\Delta E_R = 10$ meV. This rather large variation was chosen to account not only for the INS energy resolution (< 5 meV), but also for the fact that the commensurate feature may disperse somewhat with temperature (as seen, for example, in [23]). Figure 5.14 shows plots of the leading eigenvalue as a function of the coupling U_0 for both thin films above T_c. The dashed vertical lines represent the values of U_0 used in Sect. 5.3 and for which the resonance was exactly given by Eq. 5.14. The horizontal dashed lines are the resulting eigenvalues shown in Fig. 5.13. The horizontal dimension of the shaded rectangles represents the range of U_0 values for which the resonance energy moves from $E_R + \Delta E_R$ meV (lowest U_0) to $E_R - \Delta E_R$ meV (highest U_0). The vertical dimension represents the resulting uncertainty in the eigenvalue. Thus, we estimated a $15\% - 20\%$ variation in the leading eigenvalues, as represented in the error bars in Fig. 5.13.

5.5 The Coupling Energies U and U_0

Because of the relation $\lambda \sim U_0^2$, the eigenvalues are strongly dependent on the spin-fermion coupling U_0. One may wonder whether the procedure followed in this chapter is ultimately too sensitive to this coupling, which was indirectly obtained

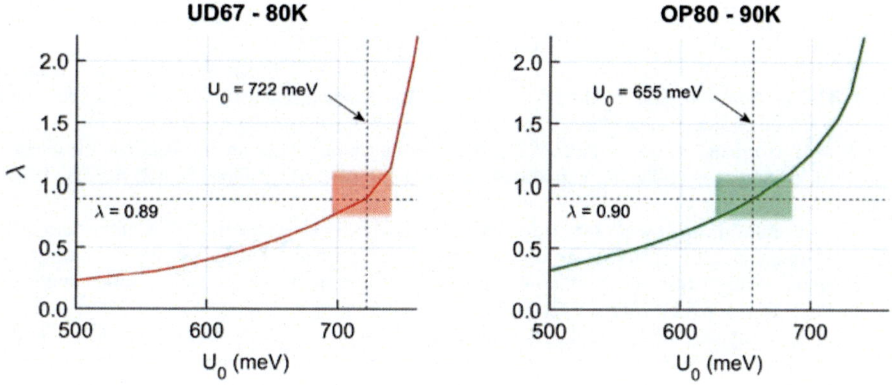

Fig. 5.14 Leading eigenvalue as a function of coupling energy for both thin films above (but close to) T_c. Shaded rectangles indicate uncertainty in U_0 and λ (reprinted from Restrepo et al. Coupled electronic and magnetic excitations in the cuprates and their role in the superconducting transition. Commun. Phys. **5**, 45 (2022))

from ARPES data. To address this question, we first note that the eigenvalues obtained by using the alternative coupling constant U (Appendix B) were comparable to the ones reported in Fig. 5.13 (compare λ_{RPA} and λ_{Σ} in Table B.1). The fact that these two independently obtained energies yielded similar eigenvalues is reassuring. Moreover, in the UD67 sample above T_c, U_0 and U take values similar to the estimate based on the spin-fermion model given in Ref. [46], $U \sim 0.7$ eV. This figure was deduced from a procedure which is formally equivalent to the RPA-based calculations presented in Sect. 5.2, and for which the free parameters were obtained from INS data on $YBa_2Cu_3O_{6.7}$ ($T_c = 67$ K) above T_c. Finally, we should also mention that the characteristic energies here obtained were somewhat lower than those inferred in other experimental works. Mishra et al. [14] and Chatterjee et al. [47], who also constructed the spin susceptibility from ARPES data, reported values of $U_0 \sim 0.9$ eV for optimally doped Bi2212, at which doping we obtained lower coupling energies (~ 0.55 eV). Dahm et al. found $U \sim 1.6$ eV for $YBa_2Cu_3O_{6.6}$ ($T_c = 61$ K). These larger energies naturally yield larger eigenvalues, leading to higher estimates of the transition temperature. Based on these comparisons, we believe the approximate transition temperatures reported here were not artificially enhanced by large U_0 or U values; in fact, the characteristic coupling energies obtained in this work were rather conservative estimates.

References

1. J.R. Schrieffer, X.G. Wen, S.C. Zhang, Dynamic spin fluctuations and the bag mechanism of high-T_c superconductivity. Phys. Rev. B **39**(16), 11663–11679 (1989)
2. A.L. Fetter, J.D. Walecka, *Quantum Theory of Many-Particle Systems* (McGraw-Hill, Boston, 1971)
3. M.R. Norman, Relation of neutron incommensurability to electronic structure in high-temperature superconductors. Phys. Rev. B **61**(21), 14751–14758 (2000)
4. M.R. Norman, Linear response theory and the universal nature of the magnetic excitation spectrum of the cuprates. Phys. Rev. B **75**(18), 184514 (2007)
5. M. Randeria, G. Jennings, H. Ding, J.C. Campuzano, A. Bellman, T. Yokoya, T. Takahashi, H. Katayama-Yoshida, T. Mochiku, K. Kadowaki, Momentum distribution sum rule for angle-resolved photoemission. Phys. Rev. Lett. **74**(24), 4951–4954 (1995)
6. M.R. Norman, M. Randeria, H. Ding, J.C. Campuzano, A.F. Bellman, Polarization selection rules and superconducting gap anisotropy in $Bi_2Sr_2CaCu_2O_8$. Phys. Rev. B **52**(21), 15107–15110 (1995)
7. A. Kaminski, S. Rosenkranz, H.M. Fretwell, J. Mesot, M. Randeria, J.C. Campuzano, M.R. Norman, Z.Z. Li, H. Raffy, T. Sato, T. Takahashi, K. Kadowaki, Identifying the background signal in angle-resolved photoemission spectra of high-temperature cuprate superconductors. Phys. Rev. B **69**(21), 212509 (2004)
8. A.A. Kordyuk, S.V. Borisenko, T.K. Kim, K.A. Nenkov, M. Knupfer, J. Fink, M.S. Golden, H. Berger, R. Follath, Origin of the Peak-Dip-Hump line shape in the superconducting-state (π,0) photoemission spectra of $Bi_2Sr_2CaCu_2O_{8+\delta}$. Phys. Rev. Lett. **89**(7), 077003
9. H. Ding, A.F. Bellman, J.C. Campuzano, M. Randeria, M.R. Norman, T. Yokoya, T. Takahashi, H. Katayama-Yoshida, T. Mochiku, K. Kadowaki, G. Jennings, G.P. Brivio, Electronic excitations in $Bi_2Sr_2CaCu_2O_{8+\delta}$: Fermi surface, dispersion, and absence of bilayer splitting. Phys. Rev. Lett. **76**(9), 1533–1536 (1996)

10. M.R. Norman, H. Ding, M. Randeria, J.C. Campuzano, T. Yokoya, T. Takeuchi, T. Takahashi, T. Mochiku, K. Kadowaki, P. Guptasarma, D.G. Hinks, Destruction of the Fermi surface in underdoped high-T_c superconductors. Nature **392**, 157 (1998)
11. A. Kanigel, U. Chatterjee, M. Randeria, M.R. Norman, G. Koren, K. Kadowaki, J.C. Campuzano, Evidence for pairing above the transition temperature of cuprate superconductors from the electronic dispersion in the pseudogap phase. Phys. Rev. Lett. **101**(13), 137002 (2008)
12. J.C. Campuzano, H. Ding, M.R. Norman, M. Randeria, A.F. Bellman, T. Yokoya, T. Takahashi, H. Katayama-Yoshida, T. Mochiku, K. Kadowaki, Direct observation of particle-hole mixing in the superconducting state by angle-resolved photoemission. Phys. Rev. B **53**(22), R14737–R14740 (1996)
13. A. Kaminski, M. Randeria, J.C. Campuzano, M.R. Norman, H. Fretwell, J. Mesot, T. Sato, T. Takahashi, K. Kadowaki, Renormalization of the spectral line shape and dispersion below T_c in $Bi_2Sr_2CaCu_2O_{8+\delta}$. Phys. Rev. Lett. **86**(6), 1070–1073 (2001)
14. V. Mishra, U. Chatterjee, J.C. Campuzano, M.R. Norman, Effect of the pseudogap on the transition temperature in the cuprates and implications for its origin. Nat. Phys. **10**, 357 (2014)
15. T. Izuyama, D.J. Kim, R. Kubo, Band theoretical interpretation of neutron diffraction phenomena in ferromagnetic metals. J. Phys. Soc. Jpn. **18**(7), 1025–1042 (1963)
16. A.V. Chubukov, D. Pines, J. Schmalian, A spin fluctuation model for d-wave superconductivity, in *Superconductivity: Conventional and Unconventional Superconductors*, ed. by K.H. Bennemann, J.B. Ketterson (Springer, Berlin, Heidelberg, 2008), pp. 1349–1413
17. J. Brinckmann, P.A. Lee, Slave Boson approach to neutron scattering in $YBa_2Cu_3O_{6+y}$ superconductors. Phys. Rev. Lett. **82**(14), 2915–2918 (1999)
18. D.Z. Liu, Y. Zha, K. Levin, Theory of neutron scattering in the normal and superconducting states of $YBa_2Cu_3O_{6+x}$. Phys. Rev. Lett. **75**(22), 4130–4133 (1995)
19. H.F. Fong, P. Bourges, Y. Sidis, L.P. Regnault, A. Ivanov, G.D. Gu, N. Koshizuka, B. Keimer, Neutron scattering from magnetic excitations in $Bi_2Sr_2CaCu_2O_{8+\delta}$. Nature **398**(6728), 588–591 (1999)
20. H.F. Fong, B. Keimer, D.L. Milius, I.A. Aksay, Superconductivity-induced anomalies in the spin excitation spectra of underdoped $YBa_2Cu_3O_{6+x}$. Phys. Rev. Lett. **78**(4), 713–716 (1997)
21. L. Capogna, B. Fauqué, Y. Sidis, C. Ulrich, P. Bourges, S. Pailhès, A. Ivanov, J.L. Tallon, B. Liang, C.T. Lin, A.I. Rykov, B. Keimer, Odd and even magnetic resonant modes in highly overdoped $Bi_2Sr_2CaCu_2O_{8+\delta}$. Phys. Rev. B **75**(6), 060502 (2007)
22. H. He, Y. Sidis, P. Bourges, G.D. Gu, A. Ivanov, N. Koshizuka, B. Liang, C.T. Lin, L.P. Regnault, E. Schoenherr, B. Keimer, Resonant spin excitation in an overdoped high temperature superconductor. Phys. Rev. Lett. **86**(8), 1610–1613 (2001)
23. P. Dai, H. A. Mook, S. M. Hayden, G. Aeppli, T. G. Perring, R. D. Hunt, and F. Doğan. The magnetic excitation spectrum and thermodynamics of high-T_c superconductors. Science **284**(5418), 1344–1347 (1999)
24. V. Hinkov, P. Bourges, S. Pailhès, Y. Sidis, A. Ivanov, C.D. Frost, T.G. Perring, C.T. Lin, D.P. Chen, B. Keimer, Spin dynamics in the pseudogap state of a high-temperature superconductor. Nat. Phys. **3**(11), 780–785 (2007)
25. B. Keimer, H.F. Fong, S.H. Lee, D.L. Millius, I.A. Aksay, Doping dependence of the magnetic resonance peak in $YBa_2Cu_3O_{6+x}$. Phys. C: Superconduct. **282–287**, 232–235 (1997)
26. G. Xu, G.D. Gu, M. Hücker, B. Fauqué, T.G. Perring, L.P. Regnault, J.M. Tranquada, Testing the itinerancy of spin dynamics in superconducting $Bi_2Sr_2CaCu_2O_{8+\delta}$. Nat. Phys. **5**(9), 642–646 (2009)
27. P.C. Hammel, M. Takigawa, R.H. Heffner, Z. Fisk, K.C. Ott, Spin dynamics at oxygen sites in $YBa_2Cu_3O_7$. Phys. Rev. Lett. **63**(18), 1992–1995 (1989)
28. A. Uldry, M. Mali, J. Roos, P.F. Meier, Anisotropy of the antiferromagnetic spin correlations in the superconducting state of $YBa_2Cu_3O_7$ and $YBa_2Cu_3O_8$. J. Phys. Condensed Matter **17**(46), L499–L505 (2005)
29. H.F. Fong, B. Keimer, D. Reznik, D.L. Milius, I.A. Aksay, Polarized and unpolarized neutron-scattering study of the dynamical spin susceptibility of $YBa_2Cu_3O_7$. Phys. Rev. B **54**(9), 6708–6720 (1996)

30. A. Kanigel, U. Chatterjee, M. Randeria, M.R. Norman, S. Souma, M. Shi, Z.Z. Li, H. Raffy, J.C. Campuzano, Protected nodes and the collapse of Fermi arcs in high-T_c cuprate superconductors. Phys. Rev. Lett. **99**(15), 157001 (2007)
31. D.K. Morr, D. Pines, The resonance peak in cuprate superconductors. Phys. Rev. Lett. **81**(5), 1086–1089 (1998)
32. N. Bulut, D.J. Scalapino, Neutron scattering from a collective spin fluctuation mode in a CuO_2 bilayer. Phys. Rev. B **53**(9), 5149–5152 (1996)
33. T. Dahm, V. Hinkov, S.V. Borisenko, A.A. Kordyuk, V.B. Zabolotnyy, J. Fink, B. Büchner, D.J. Scalapino, W. Hanke, B. Keimer, Strength of the spin-fluctuation-mediated pairing interaction in a high-temperature superconductor. Nat. Phys. **5**(3), 217–221 (2009)
34. M.K. Chan, C.J. Dorow, L. Mangin-Thro, Y. Tang, Y. Ge, M.J. Veit, G. Yu, X. Zhao, A.D. Christianson, J.T. Park, Y. Sidis, P. Steffens, D.L. Abernathy, P. Bourges, M. Greven, Commensurate antiferromagnetic excitations as a signature of the pseudogap in the tetragonal high-T_c cuprate $HgBa_2CuO_{4+\delta}$. Nat. Commun. **7**(1), 1–7 (2016)
35. S.M Hayden, G. Aeppli, P. Dai, H.A. Mook, T.G. Perring, S.W. Cheong, Z. Fisk, F. Doğan, T.E. Mason, Absolute measurements of the high-frequency magnetic dynamics in High-T_c superconductors. Phys. B: Condens. Matter **241–243**, 765–772 (1997)
36. S.M. Hayden, H.A. Mook, P. Dai, T.G. Perring, F. Doğan, The structure of the high-energy spin excitations in a high-transition-temperature superconductor. Nature **429**(6991), 531–534 (2004)
37. T.A. Maier, P. Staar, V. Mishra, U. Chatterjee, J.C. Campuzano, D.J. Scalapino, Pairing in a dry Fermi sea. Nat. Commun. **7**, 11875 (2016)
38. P.H. Monthoux, *Spin Fluctuations and High - Temperature Superconductivity in the Copper Oxides*. PhD Thesis, University of Illinois at Urbana-Champaign, 1992
39. J.E. Hirsch, Two-dimensional Hubbard model: Numerical simulation study. Phys. Rev. B **31**(7), 4403–4419 (1985)
40. A.-M.S. Tremblay, B. Kyung, D. Sénéchal, Pseudogap and high-temperature superconductivity from weak to strong coupling. Towards a quantitative theory (Review Article). Low Temp. Phys. **32**(4), 424–451 (2006)
41. T.A. Maier, M. Jarrell, D.J. Scalapino, Pairing interaction in the two-dimensional Hubbard model studied with a dynamic cluster quantum Monte Carlo approximation. Phys. Rev. B **74**(9), 094513 (2006)
42. T.A. Maier, M.S. Jarrell, D.J. Scalapino, Structure of the pairing interaction in the two-dimensional Hubbard model. Phys. Rev. Lett. **96**(4), 047005 (2006)
43. N. Bulut, D.J. Scalapino, S.R. White, Bethe-Salpeter eigenvalues and amplitudes for the half-filled two-dimensional Hubbard model. Phys. Rev. B **47**(21), 14599–14602 (1993)
44. S.R. White, D.J. Scalapino, Stripes on a 6-leg Hubbard ladder. Phys. Rev. Lett. **91**(13), 136403 (2003)
45. S. Dal Conte, L. Vidmar, D. Golež, M. Mierzejewski, G. Soavi, S. Peli, F. Banfi, G. Ferrini, R. Comin, B.M. Ludbrook, L. Chauviere, N.D. Zhigadlo, H. Eisaki, M. Greven, S. Lupi, A. Damascelli, D. Brida, M. Capone, J. Bonča, G. Cerullo, C. Giannetti, Snapshots of the retarded interaction of charge carriers with ultrafast fluctuations in cuprates. Nat. Phys. **11**(5), 421–426 (2015)
46. Ar. Abanov, A.V. Chubukov, M. Eschrig, M.R. Norman, J. Schmalian, Neutron resonance in the cuprates and its effect on fermionic excitations. Phys. Rev. Lett. **89**(17), 177002 (2002)
47. U. Chatterjee, D.K. Morr, M.R. Norman, M. Randeria, A. Kanigel, M. Shi, E. Rossi, A. Kaminski, H.M. Fretwell, S. Rosenkranz, K. Kadowaki, J.C. Campuzano, Dynamic spin-response function of the high-temperature $Bi_2Sr_2CaCu_2O_{8+\delta}$ superconductor from angle-resolved photoemission spectra. Phys. Rev. B **75**(17), 172504 (2007)

Chapter 6
Conclusions

Abstract In this last chapter we summarize the main conclusions drawn from this dissertation. The findings in this work suggest that spin fluctuations can approximately account for the experimentally observed critical temperatures and the energy and momentum dependence of the pairing interaction in the cuprates. Therefore, magnetic excitations are a plausible explanation for Cooper pairing in these materials.

As noted in the introduction, superconductivity in the cuprates emerges following quantum melting of the antiferromagnetic Mott insulating state of the parent compound, either via electron or hole doping. In this work we examined the proposal that the spin fluctuations resulting from the melted Mott state act as the "glue" leading to Cooper pairing in high temperature superconductors [1, 2]. In particular, we addressed the important question of whether the spin-fermion interaction V_{SF} is strong enough to yield the high transition temperatures observed in these materials. To this end, we considered a generalization of the BCS gap equation containing the basic ingredients for superconductivity: two electron propagators (the Green's functions G) and a pairing interaction (V_{SF}) between them. Purely theoretical approaches to this problem invariably face the challenge of obtaining reliable functional forms for these quantities. Here, we surmounted this difficulty by deriving them from ARPES data from the cuprate Bi2212. The main conclusions drawn from this work can be summarized as follows:

- Solution of the BSE required a functional form of the pairing mechanism V_{SF}. Since this function depends on the spin susceptibility, we first had to obtain this quantity from the ARPES data. Even though the calculation of $\mathrm{Im}\chi$ was forced on us by the necessity of constructing the pairing interaction, this effort also provided insights into the nature of the spin-fermion interaction in the cuprates. Focusing on the (π, π) feature, we saw that it was a broad peak above T_c and sharpened as the temperature was lowered, especially below T_c. This can be understood in the light of the spin-fermion interaction. Since a pronounced gap in the electronic excitations opens up below T_c, low-energy spin excitations

F. Restrepo, *Angle-Resolved Photoemission Spectroscopy Study of Spin Fluctuations in the Cuprate Superconductors*, Springer Theses,
https://doi.org/10.1007/978-3-031-10979-9_6

have less electrons to scatter from and are thus more sharply defined. Above T_c, the gap begins to fill-in (in the pseudogap region) and therefore the number of low-energy electronic states available for scattering increases. This explains why the resonance feature broadens and loses intensity. Furthermore, the relative insensitivity of the resonance energy to temperature was also consistent with the roughly T-independent ARPES gap. Our calculations also captured the decrease in the resonance intensity with increasing doping, a result which also tracks the doping dependence of the spectral gap seen in ARPES. Finally, we put these comparisons on a more quantitative footing by comparing the integrated magnetic spectral weights with those obtained experimentally from similarly doped YBCO samples. The overall correspondence between ARPES and INS features strongly suggests that the coupling between charge and spin degrees of freedom is an important feature of the cuprate superconductors.

- Of course, the fact that there is a spin-fermion interaction does not mean that this is the interaction responsible for superconductivity. To explore this possibility, we studied the Bethe-Salpeter equation. With ARPES-derived functions G and V_{SF}, we tracked the behavior of the eigenvalue at different temperatures and extracted the eigenfunctions near the temperatures where $\lambda \sim 1$ (signaling the onset of a superconducting transition). Our main finding was that these temperatures were in the range $60 - 90$ K, which are close to the *true* values obtained in the doping range considered here. Naturally, the eigenfunctions associated with these eigenvalues must reflect existing experimental observations. In this regard, we found these eigenfunctions had d-wave symmetry in momentum space, which is a well-known property of the cuprate order parameter. Moreover, the energy dependence of the eigenfunctions near T_c suggested that the dynamics of the pairing interaction was governed by magnetic excitations, which extend up to energies $2J$. Indeed, using experimental J values for underdoped cuprates ($\sim$ 125 meV [3]), the energy decay of the functions Φ was found to be of order $2J$, which is consistent with the time-retardation of the pairing interaction observed with ultrafast optical spectroscopy [4]. Taken together, these observations show that the spin-fluctuation-mediated pairing interaction produces (1) appropriate values of the critical temperature and (2) eigenfunctions reflecting the energy-momentum dependence of the order parameter observed in various experiments on the cuprates.

Overall, our findings suggest that the RPA and the spin fluctuation formalism may provide a consistent description of Bi2212 connecting the electronic (ARPES) and magnetic (INS) responses and yielding a high-temperature superconducting transition to a state with d-wave symmetry.

Since the calculation of the Bethe-Salpeter eigenvalues involved a series of steps and several approximations, the guiding principle of this work was to verify consistency between the intermediate results and experimental data at every step. This not only justified the approximations made, but also brought out the internal consistency of the formalism. Therefore, we hope this work will also be regarded as integrative, bringing together diverse experimental observations (as well as numerical [1] and analytical [2] calculations) under a common viewpoint, in which the interaction between electrons and antiferromagnetic excitations plays a key role in the description of the cuprate superconductors at low energies.

References

1. D.J. Scalapino, A common thread: the pairing interaction for unconventional superconductors. Rev. Mod. Phys. **84**(4), 1383–1417 (2012)
2. A.V. Chubukov, D. Pines, J. Schmalian, A spin fluctuation model for d-Wave superconductivity, in *Superconductivity: Conventional and Unconventional Superconductors*, ed. by K.H. Bennemann, J.B. Ketterson (Springer, Berlin, Heidelberg, 2008), pp. 1349–1413
3. S.M. Hayden, G. Aeppli, P. Dai, H.A. Mook, T.G. Perring, S.W. Cheong, Z. Fisk, F. Doğan, T.E. Mason, Absolute measurements of the high-frequency magnetic dynamics in high-T_c superconductors. Phys. B: Condens. Matter **241–243**, 765–772 (1997)
4. S. Dal Conte, L. Vidmar, D. Golež, M. Mierzejewski, G. Soavi, S. Peli, F. Banfi, G. Ferrini, R. Comin, B.M. Ludbrook, L. Chauviere, N.D. Zhigadlo, H. Eisaki, M. Greven, S. Lupi, A. Damascelli, D. Brida, M. Capone, J. Bonča, G. Cerullo, C. Giannetti, Snapshots of the retarded interaction of charge carriers with ultrafast fluctuations in cuprates. Nat. Phys. **11**(5), 421–426 (2015)

Appendix A
Solution of the Bethe-Salpeter Equation

This appendix covers the technical details behind the solution of the Bethe-Salpeter equation (BSE). In particular, the solution method presented here focuses on finding the leading eigenvalue and eigenfunction of the BSE, for which we employ the *power method* and the Fourier convolution theorem.

The method employed in solving the eigenvalue problem

$$\lambda(T)\Phi(T) = \hat{O}(T)\Phi(T) \tag{A.1}$$

with

$$\begin{aligned}
&\hat{O}(T)\Phi(\mathbf{k}, \omega_n, T) \\
&\quad = -\frac{T}{N}\sum_{\mathbf{k}',\Omega_n} V_{\mathrm{SF}}(\mathbf{k}-\mathbf{k}', i\omega_n - i\Omega_n, T) \\
&\qquad \times G(\mathbf{k}', i\Omega_n, T)G(-\mathbf{k}', -i\Omega_n, T)\Phi(\mathbf{k}', \Omega_n, T)
\end{aligned} \tag{A.2}$$

closely followed that in ref. [1]. This appendix covers some details of the solution of the BSE. First, we show how the convolution sums in Eq. A.2 can be simplified with Fourier transforms. Next, we introduce the iterative power method for obtaining the leading eigenvalue of the BSE. Finally, we test the algorithm with two simple eigenvalue problems where the solution is known.

A.1 Fourier Techniques for Convolution Sums

Since the power method involves performing the operation $\hat{O}\Phi$ many times (see next section), it is convenient to perform this operation in an efficient manner.

F. Restrepo, *Angle-Resolved Photoemission Spectroscopy Study of Spin Fluctuations in the Cuprate Superconductors*, Springer Theses,
https://doi.org/10.1007/978-3-031-10979-9

Defining

$$\gamma(\mathbf{k}', \Omega_n, T) \equiv G(\mathbf{k}', i\Omega_n, T)G(-\mathbf{k}', -i\Omega_n, T)\Phi(\mathbf{k}', \Omega_n, T), \tag{A.3}$$

we rewrite the BSE as

$$\lambda(T)\Phi(\mathbf{k}, \omega_n, T) = -\frac{3U_0^2}{2}\frac{1}{N\beta}\sum_{\mathbf{k}', \Omega_n} \chi(\mathbf{k} - \mathbf{k}', i\omega_n - i\Omega_n, T)\gamma(\mathbf{k}', i\Omega_n, T), \tag{A.4}$$

from which we see that the right-hand side of equation Eq. A.2 is essentially an (energy-momentum) convolution of the functions χ and γ. Such convolution sums can be easily handled thanks to the Fourier convolution theorem, as we illustrate below. We consider the momentum and energy sums separately, beginning with the momentum sum.

For now, we ignore the factor $-1.5U_0^2$ and keep only the momentum (or position) dependence of the functions. Since the functions we are dealing with are periodic in the reciprocal lattice, they can be expanded in plane waves as [2]

$$G(\mathbf{k}) = \sum_{\mathbf{R}} G(\mathbf{R})e^{i\mathbf{k}\cdot\mathbf{R}}, \tag{A.5}$$

where

$$G(\mathbf{R}) = \frac{V}{N}\int \frac{d\mathbf{k}}{(2\pi)^3} e^{-i\mathbf{k}\cdot\mathbf{R}} G(\mathbf{k}) = \frac{1}{N}\sum_{\mathbf{k}} e^{-i\mathbf{k}\cdot\mathbf{R}} G(\mathbf{k}) \tag{A.6}$$

is the Fourier coefficient of $G(\mathbf{k})$ at the Bravais lattice vector $\mathbf{R}$ and N is the number of Bravais lattice sites (and of points in the first BZ). The $\mathbf{R}$ sum runs over the N lattice sites and the $\mathbf{k}$ variable is integrated over the first BZ. With these definitions, we can write the momentum convolution as:

$$\begin{aligned} &\frac{1}{N}\sum_{\mathbf{k}'} \chi(\mathbf{k} - \mathbf{k}')\gamma(\mathbf{k}') \\ &= \frac{1}{N}\sum_{\mathbf{k}'} \Big\{\sum_{\mathbf{R}} \chi(\mathbf{R})e^{i(\mathbf{k}-\mathbf{k}')\cdot\mathbf{R}}\Big\}\Big\{\sum_{\mathbf{R}'} \gamma(\mathbf{R}')e^{i\mathbf{k}'\cdot\mathbf{R}'}\Big\}. \end{aligned} \tag{A.7}$$

Re-arranging the sums:

$$\frac{1}{N}\sum_{\mathbf{k}'}\chi(\mathbf{k}-\mathbf{k}')\gamma(\mathbf{k}')$$
$$=\frac{1}{N}\sum_{\mathbf{R}}\sum_{\mathbf{R}'}\chi(\mathbf{R})\gamma(\mathbf{R}')e^{i\mathbf{k}\cdot\mathbf{R}}\underbrace{\sum_{\mathbf{k}'}e^{i\mathbf{k}'\cdot(\mathbf{R}'-\mathbf{R})}}_{N\delta_{R,R'}}=\sum_{\mathbf{R}}\chi(\mathbf{R})\gamma(\mathbf{R})e^{i\mathbf{k}\cdot\mathbf{R}}. \quad (A.8)$$

The momentum-convolution of χ and γ is proportional to the inverse Fourier transform of the algebraic product of the Fourier transforms of the two functions. These transforms are easily calculated in technical computing software like Matlab or Mathematica. From the Matlab documentation, we see that the two-dimensional transforms `fft2` (direct) and `ifft2` (inverse) are defined by

$$\sum_{\mathbf{k}}e^{-i\mathbf{k}\cdot\mathbf{R}}F(\mathbf{k})=\mathtt{fft2}\big[F(\mathbf{k})\big] \quad \text{and} \quad \sum_{\mathbf{R}}e^{i\mathbf{k}\cdot\mathbf{R}}F(\mathbf{R})=N\mathtt{ifft2}\big[F(\mathbf{R})\big]. \quad (A.9)$$

Therefore, the momentum sum can be finally written as

$$\frac{1}{N}\sum_{\mathbf{k}'}\chi(\mathbf{k}-\mathbf{k}')\gamma(\mathbf{k}')=\frac{1}{N}\mathtt{ifft2}\big[\mathtt{fft2}(\chi_{\mathbf{k}})\mathtt{fft2}(\gamma_{\mathbf{k}})\big]. \quad (A.10)$$

The Matsubara energy convolution is handled similarly, but we start from the (discrete) definitions for the Fourier transform pairs [3]:

$$G(\omega_n)=\frac{\beta}{N_p}\sum_{\tau}G(\tau)e^{i\omega_n\tau} \quad \text{and} \quad G(\tau)=\frac{1}{\beta}\sum_{\omega_n}G(\omega_n)e^{-i\omega_n\tau}, \quad (A.11)$$

where N_p is the total number of Matsubara frequencies. With these definitions, we may write:

$$\frac{1}{\beta}\sum_{\Omega_n}\chi(\omega_n-\Omega_n)\gamma(\Omega_n)=\frac{1}{\beta}\sum_{\Omega_n}\Big\{\frac{\beta}{N_p}\sum_{\tau}\chi(\tau)e^{i\tau(\omega_n-\Omega_n)}\Big\}\Big\{\frac{\beta}{N_p}\sum_{\tau'}\gamma(\tau')e^{i\tau'\Omega_n}\Big\}. \quad (A.12)$$

Rearranging the sums, we get

$$\frac{1}{\beta}\sum_{\Omega_n}\chi(\omega_n-\Omega_n)\gamma(\Omega_n)=\frac{\beta}{N_p^2}\sum_{\tau,\tau'}\chi(\tau)\gamma(\tau')e^{i\tau\omega_n}\underbrace{\sum_{\Omega_n}e^{i\Omega_n(\tau'-\tau)}}_{N_p\delta_{\tau,\tau'}}$$
$$=\frac{\beta}{N_p}\sum_{\tau}\chi(\tau)\gamma(\tau)e^{i\tau\omega_n}. \quad (A.13)$$

Again, translating this result into Matlab language, we have

$$F(\tau) = \frac{1}{\beta}\texttt{fft}\big[F(\omega_n)\big] \qquad \text{and} \qquad F(\omega_n) = \beta\texttt{ifft}\big[F(\tau)\big]. \tag{A.14}$$

With these identifications, we may write the energy sum as

$$\frac{1}{\beta}\sum_{\Omega_n}\chi(\omega_n - \Omega_n)\gamma(\Omega_n) = \frac{1}{\beta}\texttt{ifft}\big[\texttt{fft}(\chi_{\omega_m})\texttt{fft}(\gamma_{\omega_n})\big], \tag{A.15}$$

where the spin susceptibility depends on bosonic Matsubara energies, denoted by ω_m. Going back to the original gap equation, we see that we can combine the energy and momentum transforms into a 3-D transform to get

$$\lambda\Phi = \hat{O}\Phi = -\frac{3U_0^2}{2}\frac{1}{N\beta}\texttt{ifft3}\big[\texttt{fft3}(\chi_{\mathbf{k},\omega_m})\texttt{fft3}(\gamma_{\mathbf{k},\omega_n})\big]. \tag{A.16}$$

Note that the entire operation $\hat{O}\Phi$ can now be performed in one line of code. This simplicity is afforded by the structure of the BSE in Matsubara frequencies; i.e., it is a pure convolution in all the independent variables. This simple convolution (or correlation) structure is typically lost when analytically continuing the functions to the real energy axis. A case in point is the bare susceptibility calculated in Sect. 5.2 (Eq. 5.9), for which the energy integrals had to be performed explicitly, with various nested *for* loops. Aside from simplicity, expressing convolution (or correlation) sums in terms of Fourier transforms results in much shorter computation times, since the operations `fft` and `ifft` are the Matlab implementations of the highly efficient *Fast Fourier Transform* algorithm [4]. This was the main reason for solving the pairing problem in Matsubara formalism.

A.2 The Power Method

In assessing whether a hypothetical pairing interaction is strong enough to mediate superconductivity at a given temperature, it is sufficient to focus on the largest eigenvalue of $\hat{O}$. The *power method* is designed precisely for this purpose. There are several equivalent ways of carrying out this procedure; in this work, the implementation goes as follows:

- We start with a guess for the eigenfunction Φ, which is a generalized gap function reflecting the momentum and energy dependence of the pairing interaction. Motivated by the well-established d-wave form of this interaction, we take $\Phi_0(\mathbf{k}, \omega_n) = \cos(k_x) - \cos(k_y)$ as our starting guess. From this initial choice for the eigenfunction Φ_0 we form an initial guess γ_0 using Eq. A.3.

- With this guess, we perform the first iteration $F_1 = \hat{O}\Phi_0$ (using Eq. A.16) and factor out the largest entry in F_1, so that $F_1 = \lambda_1\Phi_1$, where $\lambda_1 = \max(F_1)$ and $\Phi_1 = \frac{F_1}{\lambda_1}$.
- From Φ_1, we form γ_1 (according to Eq. A.3) and perform the second iteration $F_2 = \hat{O}\Phi_1$ to obtain $\lambda_2 = \max(F_2)$ and $\Phi_2 = \frac{F_2}{\lambda_2}$, as before.
- The above process is repeated until the iteration (N) is reached where the difference $\lambda_N - \lambda_{N-1}$ is negligible. At this point, the function Φ_N is the eigenfunction of $\hat{O}$ corresponding to the largest eigenvalue λ_N.

From the Lehmann representation of the Green's function, it can be seen that $P(k, \omega_n) = G(k, \omega_n)G(-k, -\omega_n)$ is even in Matsubara frequency, real, and positive. For the Matsubara spin response, the property $\text{Im}\chi(q, -\Omega) = -\text{Im}\chi(q, \Omega)$ implies that $\chi(q, \omega_m)$ is also even in ω_m, real and positive. These properties are not only useful as consistency checks of the numerical procedure, but also enable use of the power method, which breaks down when the eigenvalues are complex.

A.3 Simple Tests of the Algorithm for the BSE

In this section we present two examples of convolution equations which are one-dimensional analogues of the BSE considered in this work. We begin with an equation frequently found in signal processing applications:

$$\int_{-\infty}^{\infty} \text{sinc}(\omega - \omega')\psi(\omega')d\omega' = \lambda\psi(\omega). \tag{A.17}$$

The convolution in Equation Eq. A.17 can be expressed as

$$F^{-1}\Big\{F[\text{sinc}(\omega)]F[\psi(\omega)]\Big\}, \tag{A.18}$$

where F and F^{-1} denote Fourier and inverse Fourier transforms, respectively. Since $F[\text{sinc}(\omega)]$ is just a "boxcar" function, if $F[\psi(\omega)]$ is narrower than this function, then $F[\text{sinc}(\omega)]F[\psi(\omega)] \sim F[\psi(\omega)]$. This situation corresponds to a band pass filter which is broader than the input signal, in which case the signal is passed unaffected. Therefore, we expect $\psi(x)$ to be an eigenfunction of the sinc-convolution operator with $\lambda = 1$. To test this, we choose a Gaussian test function $\psi_0(x) = e^{-0.1x^2}$ for our initial guess and set $\lambda_0 = 0.5$ initially. Because in this simple case the initial guess *is* the eigenfunction, the power method immediately converges to the eigenfunction $\psi = \psi_0$ with $\lambda = 1$, as shown in Fig. A.1.

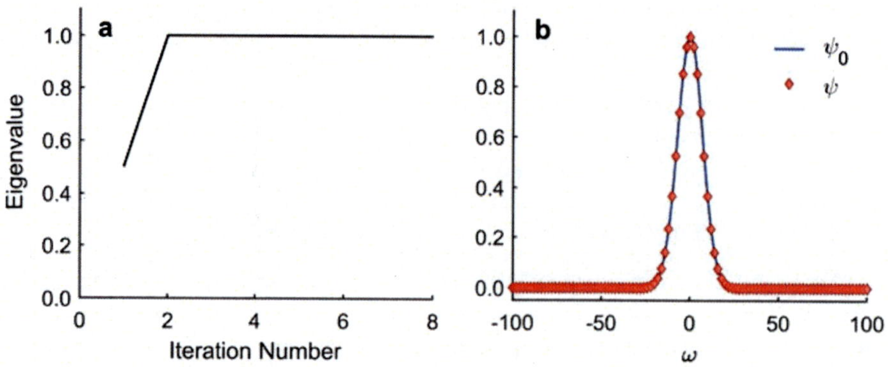

Fig. A.1 (**a**) Behavior of the eigenvalue as a function of iteration number in the power method for the sinc-convolution problem. (**b**) Initial eigenfunction guess (blue line) and calculated eigenfunction (red diamonds)

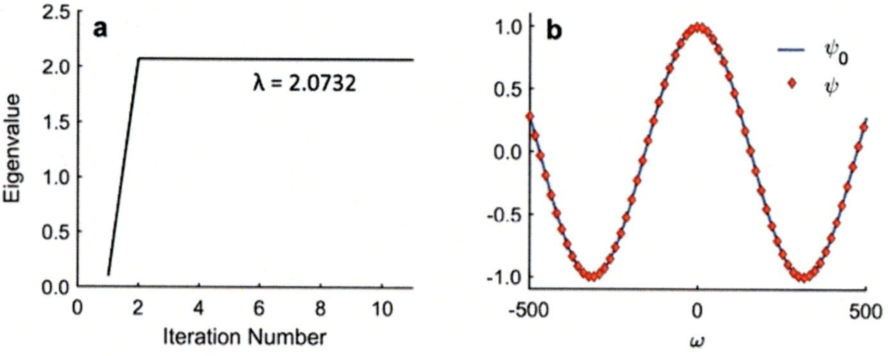

Fig. A.2 (**a**) Behavior of the eigenvalue as a function of iteration number in the power method. (**b**) Initial eigenfunction guess (blue line) and calculated eigenfunction (red diamonds)

As a second example, consider another convolution product $f(\omega) * \psi(\omega)$, where $\psi(\omega) = e^{\alpha\omega}$ (α is a complex number). It can be shown that

$$\int_{-\infty}^{\infty} f(\omega')\psi(\omega - \omega')d\omega' = \lambda_\alpha \psi(\omega), \tag{A.19}$$

with

$$\lambda_\alpha = \int_{-\infty}^{\infty} f(\omega)e^{-\alpha\omega}d\omega \tag{A.20}$$

Thus, $\psi(\omega)$ is the eigenfunction of a convolution operator (with $f(\omega)$ as a kernel) with an eigenvalue given by Eq. A.20. To test our procedure, we take a gaussian "test" kernel $f(x) = e^{-\omega^2}$, and assume $\alpha = 0.01i$ (this small value was chosen so that the imaginary components in Eq. A.19 vanish). The eigenvalue obtained from Eq. A.20 is 2.0732. To apply the power method, we take $\psi_0(\omega) = e^{(0.01i)\omega}$ and $\lambda_0 = 0.1$ as our starting guesses and compute the convolution integral with Fourier transforms (as done with the BSE in this work). Figure A.2a shows the eigenvalue as a function of iteration number. Again, we see that λ converges very quickly to the expected 2.0732. Moreover, Fig. A.2b shows that the calculated eigenfunction again matches the initial guess $\psi_0(\omega)$.

References

1. P.H. Monthoux, *Spin Fluctuations and High - Temperature Superconductivity in the Copper Oxides*. PhD Thesis, University of Illinois at Urbana-Champaign, 1992
2. N.W. Ashcroft, N.D. Mermin, *Solid State Physics* (Brooks/Cole, Monterey, 1976)
3. A.L. Fetter, J.D. Walecka, *Quantum Theory of Many-Particle Systems* (McGraw-Hill, Boston, 1971)
4. M. Frigo, S.G. Johnson, The design and implementation of FFTW3. Proc. IEEE **93**(2), 216–231 (2005)

Appendix B
Alternative Estimate of the Spin-Fermion Coupling from the Single-Particle Self-energy

In this appendix, an alternative estimate for the spin-fermion coupling, based on the renormalization of the Fermi velocity, is presented. These estimates yield results which are in quantitative agreement with those obtained in the main text.

It has been pointed in Ref. [1] that the the effective electron-electron interaction should be written as

$$V(\mathbf{q}, \Omega) = \frac{3}{2} U^2 \frac{\chi_0(\mathbf{q}, \Omega)}{1 - U_q \chi_0(\mathbf{q}, \Omega)}, \tag{B.1}$$

where the coupling energy U may differ from U_0 due to vertex corrections. Following ref. [1], here we show how this energy can be obtained from the electron self-energy. In this section, we only report results in the important region near T_c.

Assuming the single-particle renormalizations are due to V_{SF}, we can estimate the self-energy from the measured spectral functions and the spin susceptibilities derived in Sects. 5.1 and 5.2:

$$\text{Im}\Sigma(\mathbf{k}, \omega) = \frac{1}{N} \frac{3}{2} U^2 \sum_q \int d\Omega [n(\Omega) + f(\Omega - \omega)] \text{Im}\chi(\mathbf{q}, \Omega) A(\mathbf{k} - \mathbf{q}, \omega - \Omega), \tag{B.2}$$

where $n(\omega)$ and $f(\omega)$ are the Bose and Fermi distributions, respectively. For the energy integral, we used the cut-off condition $\text{Im}\chi(\mathbf{q}, \Omega) = 0$ for $\Omega > 0.4$ eV. The momentum sum in Eq. B.2 is also seen to have the form of a convolution, so we used Fourier techniques to speed its computation. The energy integral is not immediately amenable to such treatment; however, note that the integrand can be separated as (neglecting the momentum variable temporarily)

$$n(\Omega)\text{Im}\chi(\Omega)A(\omega - \Omega) + \text{Im}\chi(\Omega)A(\omega - \Omega) - f(\omega - \Omega)\text{Im}\chi(\Omega)A(\omega - \Omega),$$

F. Restrepo, *Angle-Resolved Photoemission Spectroscopy Study of Spin Fluctuations in the Cuprate Superconductors*, Springer Theses,
https://doi.org/10.1007/978-3-031-10979-9

Table B.1 Comparison between the coupling energies U_0 and U and their corresponding eigenvalues (λ_{RPA} and λ_Σ, respectively) (Reprinted from Restrepo et al. Coupled electronic and magnetic excitations in the cuprates and their role in the superconducting transition. Commun. Phys. **5**, 45 (2022))

Sample	T (K)	U_0 (meV)	U (meV)	λ_{RPA}	λ_Σ
UD67	80	722	775	0.9	1.02
	62	741	820	0.87	1.1
OP80	90	655	645	0.92	0.86
	70	660	635	1.01	0.93
OP91	90	585	525	1.05	0.84

where all terms are of the form $f(\Omega)g(\omega - \Omega)$, so that their integrals are also convolutions. By Kramers-Kronig relations, we can then obtain the real part of the self-energy and the nodal Fermi velocity renormalization,

$$Z = \frac{v_F^0}{v_F} = 1 - \frac{d}{d\omega}\mathrm{Re}\Sigma(k_n, \omega)\bigg|_{\omega=0}, \tag{B.3}$$

where k_n is the nodal Fermi momentum and v_F^0 and v_F are the bare and interacting nodal Fermi velocities, respectively. Based on the T-linear behavior of this velocity inferred from laser ARPES data on optimally doped Bi2212 [2], we estimate that, for the OP80 sample, Z(70K) $\sim$ 2.1 and Z(90K) $\sim$ 2, with similar values for the single crystal OP91. For the UD67, we note that nodal quasiparticle renormalizations also increase with underdoping (as shown in [3]), from which we estimate Z(62K) $\sim$ 2.4 and Z(80K) $\sim$ 2.3. Using the measured spectral functions and the spin susceptibilities calculated in the previous sections, we computed Z while adjusting U until we obtained the values above. In all cases, U and U_0 differ by at most 11%, which is somewhat lower than the deviation reported in [1]. Using the latter couplings we repeated the calculations for the BSE and the leading eigenvalues. The results, tabulated in the last column of Table B.1 (λ_Σ), show that the eigenvalues are still essentially equal to unity near T_c. The λ_{RPA} values are the same as those plotted in Fig. 5.13 and are included here for comparison.

References

1. V. Mishra, U. Chatterjee, J.C. Campuzano, M.R. Norman, Effect of the pseudogap on the transition temperature in the cuprates and implications for its origin. Nat. Phys. **10**, 357 (2014)
2. N.C. Plumb, T.J. Reber, J.D. Koralek, Z. Sun, J.F. Douglas, Y. Aiura, K. Oka, H. Eisaki, D.S. Dessau, Low-energy (< 10 meV) feature in the nodal electron self-energy and strong temperature dependence of the Fermi velocity in $Bi_2Sr_2CaCu_2O_{8+\delta}$. Phys. Rev. Lett. **105**(4), 046402 (2010)

3. I.M. Vishik, W.S. Lee, F. Schmitt, B. Moritz, T. Sasagawa, S. Uchida, K. Fujita, S. Ishida, C. Zhang, T.P. Devereaux, Z.X. Shen, Doping-dependent nodal Fermi velocity of the high-temperature superconductor $Bi_2Sr_2CaCu_2O_{8+\delta}$ revealed using high-resolution angle-resolved photoemission spectroscopy. Phys. Rev. Lett. **104**(20), 207002 (2010)

Appendix C
Additional Results for the OP91 Sample

This appendix presents additional results obtained from ARPES data from the optimally doped single crystal (OP91), which were omitted in the main text to avoid redundancy. We present the calculated spin susceptibility, as well as the energy and momentum dependence of the pairing eigenfunction.

In Sect. 5.1 we showed an example of the OP91 data used in the calculation of the spin susceptibility and the solution of the BSE. For these data, however, we did not show our results for Imχ nor the eigenfunction Φ, to avoid redundancy. Figure C.1 summarizes the results for the OP91 crystal, emphasizing those obtained at 90 K ($\sim T_c$). In real frequencies, Imχ again broadens and loses intensity as the temperature is increased (Fig. C.1a), while the upward dispersion branch is still visible in Fig. C.1b. Pink symbols in this panel represent INS data from optimally doped Bi2212 at 100 K [1]. Since the obtained Fermi surface for the OP91 crystal is flatter around the node than for the thin films, enhanced FS nesting in the former gives rise to stronger linear dispersing features on either side of the (π, π) feature (compare panel b with Fig. 5.9). The pairing eigenfunction Φ also changes sign upon crossing the BZ diagonal and decays with a similar energy scale as in the thin films ($\sim$ 250 meV), as shown in Fig. C.1c, d.

F. Restrepo, *Angle-Resolved Photoemission Spectroscopy Study of Spin Fluctuations in the Cuprate Superconductors*, Springer Theses,
https://doi.org/10.1007/978-3-031-10979-9

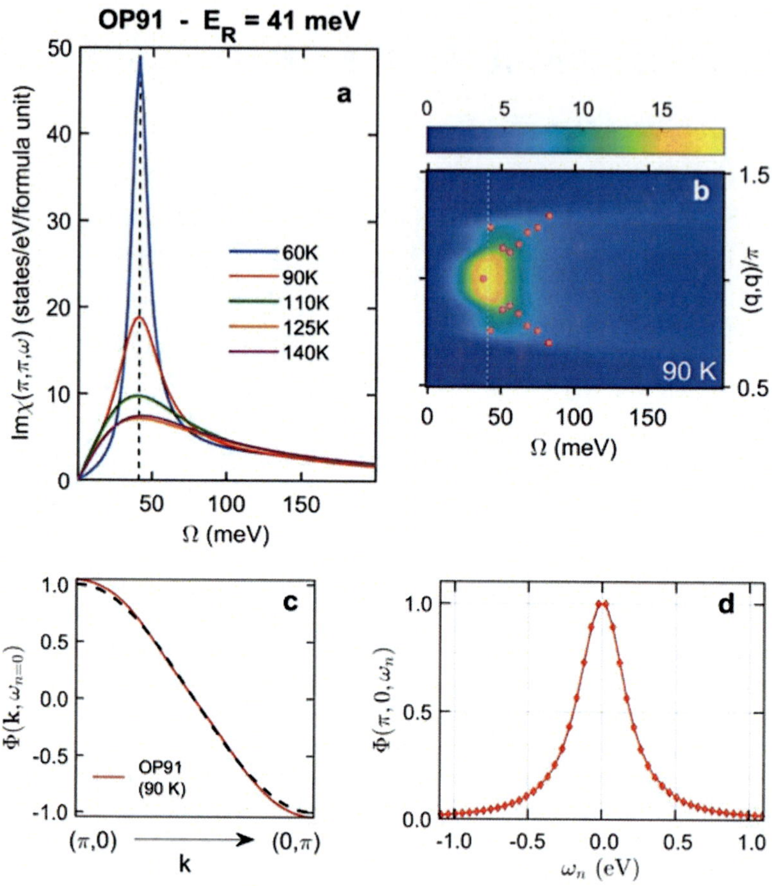

Fig. C.1 Additional calculations for the OP91 Bi2212 single crystal. (**a**) ARPES-derived spin susceptibility at the commensurate wave-vector (π, π) for temperatures between 60 K (below T_c) and 140 K ($\sim T^*$). (**b**) Energy-momentum dispersion of the spin susceptibility along the $(\pi/2, \pi/2) - (3\pi/2, 3\pi/2)$ direction. Pink symbols represent INS measurements (only above the resonance energy) from optimally doped Bi2212 at 100 K [1]. (**c**) Generalized order parameter for 90 K data along $(\pi, 0) - (0, \pi)$ direction (dashed line is the pure d-wave function $\cos k_x - \cos k_y$). (**d**) Matsubara energy dependence of order parameter at $(\pi, 0)$ at 90 K (Reprinted from Restrepo et al. Coupled electronic and magnetic excitations in the cuprates and their role in the superconducting transition. Commun. Phys. **5**, 45 (2022))

Reference

1. G. Xu, G.D. Gu, M. Hücker, B. Fauqué, T.G. Perring, L.P. Regnault, J.M. Tranquada, Testing the itinerancy of spin dynamics in superconducting $Bi_2Sr_2CaCu_2O_{8+\delta}$. Nat. Phys. **5**(9), 642–646 (2009)

Appendix D
The Anomalous Green's Function

Two-particle correlation functions in the superconducting state are typically expressed as integrals involving the normal and anomalous Green's functions. This appendix justifies neglect of the anomalous Green's function throughout the calculations presented in this work.

In the superconducting state, the particle-hole bubble χ_0 should include a term quadratic in the anomalous Green's function, which we call FF term. We ignored this term in this work for three reasons:

- As pointed out by Chatterjee et al. [1], in the superconducting state, the calculated dynamic susceptibilities vary only slightly when one includes the anomalous Green's function. Inclusion of the FF term will certainly influence the *bare* susceptibility, but an appropriate readjustment of the coupling constant U_0 will keep the resonance in the interacting susceptibility at the correct energy.
- While we considered several data sets below T_c to illustrate the performance of the RPA in constructing the spin fluctuation propagator, for the calculation of the Bethe-Salpeter eigenvalue we restricted our analysis to temperatures near and above T_c, where the FF term is not important.
- Similar results for the eigenvalues were obtained from an independent method of obtaining the prefactor U^2 in the pairing interaction V_{SF}, as shown in Appendix B.

However, it is instructive to examine the effect of this term on the RPA energies U_0 in the superconducting state. To this end, we first expressed the real part of the bare susceptibility as $\mathrm{Re}(\chi_0) = \mathrm{Re}(\chi_0^G) + \mathrm{Re}(\chi_0^F)$ (where the second term is the anomalous contribution coming from the FF term) and defined the ratio $\alpha = \frac{\mathrm{Re}(\chi_0^F)}{\mathrm{Re}(\chi_0^G)}$ [1]. Since χ_0^F is unknown, so is α. However, in a d-wave superconductor, $\mathrm{Re}(\chi_0^F)$ and $\mathrm{Re}(\chi_0^G)$ add constructively at the commensurate wave-vector, so we expect this ratio to be positive at (π, π). From the pole condition for the resonance, $1 - U_0\mathrm{Re}(\chi_0) = 0$, we see that an enhancement of the real part of the bare

F. Restrepo, *Angle-Resolved Photoemission Spectroscopy Study of Spin Fluctuations in the Cuprate Superconductors*, Springer Theses,
https://doi.org/10.1007/978-3-031-10979-9

susceptibility reduces U_0. To estimate this reduction, we computed $\mathrm{Re}(\chi_0^G)$ as in the main text (using ARPES data below T_c) and expressed the real part of the *total* bare susceptibility as

$$\mathrm{Re}(\chi_0) = \mathrm{Re}(\chi_0^G)(1 + \alpha_{\mathrm{BCS}}),$$

where the ratio α was calculated using BCS normal and anomalous spectral functions [2] and a tight-binding fit for the Bi2212 dispersion ϵ_k [3]. We considered superconducting state spectral functions with a Lorentzian lineshape, sampled over a momentum grid of 128×128 points and energies ranging from -1.5 to 1.5 eV. We also allowed for a finite width (Γ) of the spectral function and studied the effect of varying this width. The convergence factor δ was set to 10 meV, and superconductivity was included through the d-wave gap

$$\Delta(\mathbf{k}) = \frac{\Delta_0}{2}(\cos k_x - \cos k_y), \tag{D.1}$$

with $\Delta_0 = 35$ meV. Since we were only interested in the effect on the coupling constant U_0, we compared results only at the commensurate wave-vector (π, π). For $\Gamma = 40$ meV, $\alpha = 0.05$, leading to a 5% decrease in U_0. For $\Gamma = 20$ meV, $\alpha = 0.10$, leading to a 10% decrease in U_0. Finally, for $\Gamma = 0$ meV, $\alpha = 0.2$, leading to a 20% decrease in U_0. The values of U_0 below T_c reported in Fig. 5.10b correspond to the more physically relevant case of $\Gamma = 20$ meV. The "error" bars simply account for the range of U_0 values obtained from assuming $\Gamma = 40$ meV (highest U_0) and $\Gamma = 0$ meV (lowest U_0). We caution that these rough estimates were obtained from Lorentzian spectral functions, which are not representative of the cuprates, where self-energy effects give rise to more complicated lineshapes.

References

1. U. Chatterjee, D.K. Morr, M.R. Norman, M. Randeria, A. Kanigel, M. Shi, E. Rossi, A. Kaminski, H.M. Fretwell, S. Rosenkranz, K. Kadowaki, J.C. Campuzano, Dynamic spin-response function of the high-temperature $Bi_2Sr_2CaCu_2O_{8+\delta}$ superconductor from angle-resolved photoemission spectra. Phys. Rev. B **75**(17), 172504 (2007)
2. A.L. Fetter, J.D. Walecka, *Quantum Theory of Many-Particle Systems* (McGraw-Hill, Boston, 1971)
3. M.R. Norman, Linear response theory and the universal nature of the magnetic excitation spectrum of the cuprates. Phys. Rev. B **75**(18), 184514 (2007)